Amorphous and Microcrystalline Silicon Solar Cells: Modeling, Materials and Device Technology

Amorphous and Microcrystalline Silicon Solar Cells: Modeling, Materials and Device Technology

By

Ruud E.I. Schropp
Utrecht University, Debye Institute

Miro Zeman
Delft University of Technology

SPRINGER SCIENCE+BUSINESS MEDIA, LLC

Library of Congress Cataloging-in-Publication Data

Schropp, Ruud E. I. , 59-
Amorphous and microcrystalline silicon solar cells: modeling materials and device technology / by Ruud E.I. Schropp, Miro Zeman.
p. cm.
Includes bibliographical references and index.
ISBN 978-1-4613-7576-0 ISBN 978-1-4615-5631-2 (eBook)
DOI 10.1007/978-1-4615-5631-2
1. Solar cells -- Materials. 2. Silicon crystals. 3. Thin film devices--Design and construction. 4. Amorphous semiconductors.
I. Zeman, Miro, 1957 II. Title.
TK2960.S37 1998
621.31'244--dc21

98-42603
CIP

Originally published by Kluwer Academic Publishers in 1998
Softcover reprint of the hardcover 1st edition 1998

Printed on acid-free paper.

Contents

List of Figures

List of Tables

Preface

Amorphous silicon solar cell technology has evolved considerably since the first amorphous silicon solar cells were made at RCA Laboratories in 1974. Scientists working in a number of laboratories worldwide have developed improved alloys based on hydrogenated amorphous silicon and microcrystalline silicon. Other scientists have developed new methods for growing these thin films while yet others have developed new photovoltaic (PV) device structures with improved conversion efficiencies. In the last two years, several companies have constructed multi-megawatt manufacturing plants that can produce large-area, multijunction amorphous silicon PV modules. A growing number of people believe that thin-film photovoltaics will be integrated into buildings on a large scale in the next few decades and will be able to make a major contribution to the world's energy needs.

In this book, Ruud E. I. Schropp and Miro Zeman provide an authoritative overview of the current status of thin film solar cells based on amorphous and microcrystalline silicon. They review the significant developments that have occurred during the evolution of the technology and also discuss the most important recent innovations in the deposition of the materials, the understanding of the physics, and the fabrication and modeling of the devices.

DAVID E. CARLSON

This book is dedicated to all
who devote their time and
effort to promoting a clean and
sustainable energy source for
mankind

INTRODUCTION

Ruud Schropp

Electricity is an energy source that greatly contributes to the quality of life. It makes our household appliances work, it enhances our communication facilities, it powers electric public transportation systems, it accelerates our information retrieval and processing capabilities. Roughly two thirds of the world's electricity needs are generated by burning fossil fuels. Apart from the fact that this source will become exhausted, the detrimental environmental effects demand that cleaner, sustainable electricity sources are developed and expanded. Within the variety of sustainable energy sources, photovoltaic energy will play a considerable role. The beauty of it is that it can be applied in systems of any scale, from the mW to MW ranges, so that in principle nobody on this planet would have to be fully dependent on centrally organized power plants. Photovoltaic panels should be made of abundant, cheap, and non-toxic materials, while these materials should allow handling and processing such that manufacturing at low cost is feasible. Silicon thin film photovoltaic technology offers these prerequisites.

Since the first amorphous silicon solar cells were prepared in 1974 by David E. Carlson, the simple low temperature thin film formation process and the feasibility of coating extremely large areas have stimulated entrepreneurial spirits to start production plants of various sizes. Although photovoltaic sales have been increasing at a 15-20 % annual growth rate, the low dollar rate at which fossil fuels are available makes it hard to compete economically. Political appreciation of the avoided social costs of fossil fuel consumption would help to reach the break-even point much sooner. Meanwhile researchers around the world continue to work hard to further improve the efficiency and reliability of solar cells, with the help of the physical insight obtained by computer modeling, and to develop yet cheaper manufacturing methods.

In this book, we chose to study the most promising cross section of the wide variety of deposition technologies, the many kinds of silicon-hydrogen thin films, the fundamentally different solar cell structures, and the multidisciplinary physics and technology attributes that are presently available. The book is by no means complete; neither is the development of the ultimate thin film solar cell. Therefore its primary aim is to encourage scientists and industrialists to promote the introduction of photovoltaic energy by pushing the technology further.

In our personal quest for a breakthrough technology in the field of photovoltaic energy conversion, which has recently gained momentum, Miro Zeman and I have interacted with many researchers, industrialists, and project monitoring authorities. Therefore, first of all, we want to express our gratitude to all who showed their confidence in what we have been trying to achieve, by actively cooperating or by directly supporting our research. This research requires the long term commitment that society can provide through government institutions such as our Universities, NOVEM and NWO in the Netherlands, and the European Union RTD programmes. We have also benefited from the exchange of insights and results through scientific collaboration world wide, as well as close to home, with our nearest colleagues. It is impossible to adequately thank every individual, who contributed to our enterprises. Of the many interactions that have taken place, we wish to acknowledge Utrecht University, Delft University of Technology, Eindhoven University of Technology, TNO Institute of Applied Physics, National Renewable Energy Laboratory, Asahi Glass Co., Neste Corporation, Microchemistry Ltd., Naps France, Glasstech Solar, MVSystems, Elettrorava, FZ Jülich, Ecole Polytechnique Palaiseau, IMT Neuchâtel, and Philips Research Laboratories Redhill and Eindhoven for their scientific contributions. We also wish to acknowledge the recent cooperation with our partners at Akzo Nobel. My co-author Miro Zeman wishes to express his thankfulness to his parents for providing him continuous love and care that gave him peace and confidence for this challenging work. The greatest support for me came from my wife Mechtild, my son Floris and my daughter Gwendolijn: "Thank you for understanding what I want to accomplish in life and for helping me in every aspect".

I Technology of Amorphous and Microcrystalline Silicon Solar Cells

1 INTRODUCTION

On the basis of the results (...) we should like to conclude that, contrary to earlier suggestions, it is possible to incorporate substitutionally both pentavalent and trivalent impurities into a tetrahedral amorphous semiconductor such as Si.
—Walter Spear and Peter LeComber, 1975

1.1 HISTORY OF AMORPHOUS AND MICROCRYSTALLINE SILICON

The research on non-crystalline and heterogeneous materials has grown over the past 20 years to one of the most active areas in solid state physics. In this field, silicon has been a model material, as it can be modified from its purest, single crystalline state, via a two-phase microcrystalline or nanocrystalline state, to an almost perfectly disordered, amorphous state. The technological potential of each form of thin film silicon is tremendous, as electronic properties that are suitable for making devices can be preserved by incorporating hydrogen into the network with an appropriate concentration and bonding structure. The unique properties of amorphous and microcrystalline silicon, together with the modern techniques for preparing thin films over large areas, open many opportunities for semiconductor device applications.

Sterling and Swann, 1965 were the first to publish the formation of films of "silicon from silane" in a radio frequency glow discharge. The amorphous layers showed an extremely high resistivity. Later, Chittick et al., 1969 prepared amorphous silicon in the same manner and found that it had better photoconductive properties than that made by the traditional techniques, evaporation and sputtering. Their paper also contains the first report on the possibility of doping the material substitutionally. In the same years, also the first low-temperature microcrystalline silicon layers were deposited on glass substrates by Vepřek and Mareček, 1968, who used a hydrogen plasma and a chemical transport method at 600 °C.

It was demonstrated from field-effect measurements by Spear and LeComber, 1972 that plasma-deposited amorphous material can be made essentially with a low density of states in the semiconductor band gap. It took three more years until Spear and LeComber, 1975 and 1976, demonstrated that the material can be doped n-type and p-type over a large range by adding phosphine (PH_3) or diborane (B_2H_6) to the glow-discharge gas mixture. At around the same time, Paul et al., 1976 showed that amorphous silicon with similar doping susceptibility could be prepared by reactive sputtering in the presence of hydrogen. Also that year, Carlson and Wronski, 1976 had achieved similar doping effects, and in the same publication they reported on a thin-film amorphous silicon solar cell with an energy conversion efficiency of 2.4 %.

Until that time, it was generally thought that amorphous silicon could not be doped, as all impurity atoms would be incorporated according to their natural coordination number. In fact, it was quite challenging to explain that amorphous silicon, without translational symmetry or Brillouin zones could be doped at all. The fact that hydrogen played an important role in the newly made doped films was not recognized immediately. Shortly after Lewis et al, 1974 showed that the electronic defect density in amorphous germanium prepared by sputtering could be reduced by adding hydrogen to the plasma, Triska et al., 1975 showed by means of an evolution experiment that also amorphous silicon deposited by glow discharge from pure silane did contain hydrogen. Thus electronic-grade amorphous silicon is in fact an alloy of silicon and hydrogen, and was since then called hydrogenated amorphous silicon (a-Si:H).

The succesful doping of amorphous silicon triggered a tremendous interest in this material and the research activities in this field around the world grew explosively. It also rekindled the research on microcrystalline silicon, as Japanese groups found themselves preparing microcrystalline films instead of amorphous films in the course of a study on the preparation of doped a-Si:H by glow-discharge deposition (Usui and Kikuchi, 1979, Matsuda et al., 1980). These microcrystalline silicon (μc-Si:H) films were typically optimized in a high discharge power, high hydrogen dilution regime of the glow-discharge parameter space (Matsuda, 1983). A breakthrough in this field occurred when Hattori et al., 1987 applied highly conductive doped microcrystalline films in p-i-n solar cells. Only recently, also *intrinsic* microcrystalline films found an application in solar cells, when both the glow-discharge technique in a frequency range of 50 - 120 MHz (Faraji et al., 1992, Very High Frequency Chemical Vapor Deposition, VHFCVD) and the hot-wire decomposition technique (Hot Wire Chemical Vapor Deposition, HWCVD) yielded device-quality microcrystalline films (Rath et al., 1996). Under certain circumstances, the amorphous phase in these microcrystalline films can be completely eliminated. In this case the materials are rather referred to as *polycrystalline* thin films. Table 1.1 shows the definitions used throughout this book.

Table 1.1. Definitions for various morphologies of thin film silicon materials as used in this book.

Identification	*Symbol*	*Phases*	*Feature size*
Hydrogenated amorphous silicon	a-Si:H	single phase amorphous	none
Hydrogenated microcrystalline silicon	μc-Si:H	two-phase amorphous and crystalline	$<$ 20 nm crystals
Hydrogenated polycrystalline silicon	poly-Si:H	single phase crystalline with grain boundaries	$>$ 20 nm for the smallest crystal dimension

1.2 APPLICATIONS

Amorphous and microcrystalline silicon thin films are currently applied in solar cells and numerous other electronic devices. A very successful device is the thin film transistor (TFT), a switching device that has already found application in commercially available liquid crystal displays (LCDs), image sensors arrays (1-dimensional and 2-dimensional), and printing arrays.

Many applications benefit specifically from the large area manufacturability of these thin films. Moreover, the low processing temperatures for the formation of these high quality thin film semiconductor materials facilitate the cost-effective production of large-area devices such as complete matrices of switching devices on glass substrates. Amorphous silicon TFT-driven active matrices of LCDs can be found in pocket televisions, laptop computers, and in compact video projection systems. In linear and 2-dimensional image sensor arrays, the TFTs can be combined with photodiodes or photoconductors that are also made of amorphous silicon, thus utilizing the characteristically high photoconductivity. The sensors can be brought in direct contact with the document to be scanned and thus the bulky imaging optics can be eliminated. In page-wide printer heads, the TFTs are used to address arrays of output transducers for printing and copying. Charge-coupled devices (CCDs) are very attractive for facsimile transmitters (fax-machines) and can also be made of a-Si:H.

Since 1984, a-Si:H is used as the photoreceptor in electrophotographic drums. Here, the low dark conductivity and high photoresponse, along with the mechanical hardness and chemical inertness of a-Si:H are put to practice in the demanding commercial copying application.

The sensing capability of amorphous silicon devices can be further extended to ultraviolet light, particles, γ-rays, and x-rays, by utilizing alloys, thick p-i-n diodes, or radiation converters such as CsI in front of the diode, respectively. The latter device is finding applications in radiotherapy and radiography (Antonuk, 1998).

Microcrystalline silicon layers are often combined with amorphous layers in TFTs and solar cells, but also find dedicated applications in, for example, emitters of heterojunction bipolar transistors and low-temperature epitaxial layers on c-Si. Micro- or polycrystalline silicon can also be used as the active layer in TFTs, potentially showing higher mobility and better high frequency response. Completely micro- or polycrystalline thin film silicon solar cells have recently been obtained and receive considerable interest due to their specific spectral response and improved stability compared to a-Si:H solar cells.

Diode structures similar to solar cells potentially make efficient light-emitting devices (LEDs) due to their electroluminescent properties. For visible light-emitting thin film devices wide band gap alloys, such as a-SiC:H have been used. The p-i-n type diodes can be matrix addressed by TFTs to make a flat panel display, similar to LCD panels.

Of the many potential applications of a-Si:H and μc-Si:H we here briefly mention strain gauges, pressure sensors, color sensors, position-sensitive detectors, electrostatic loudspeakers, vidicons, memory devices, artifical neural networks, etc. For further reading on many of these applications we refer to an edited book on amorphous and microcrystalline semiconductor devices by Kanicki, 1991.

References

Antonuk, L., *Active matrix flat-panel imagers (AMFPIs) - A potentially dominant technology for medical X-ray imaging in the 21st century*, in: Amorphous and Microcrystalline Silicon Technology - 1998, edited by R. Schropp, H. Branz, S. Wagner, M. Hack, and I. Shimizu, Materials Research Society Symp. Proc. **507** (1998) in print.

Carlson, D.E. and C.R. Wronksi, *Amorphous silicon solar cell*, Appl. Phys. Lett. **28** (1976) 671-673.

Chittick, R.C., J.H. Alexander, and H.F. Sterling, *The preparation and properties of amorphous silicon*, J. Electrochem. Soc. **116** (1969) 77-81.

Faraji, M., S. Gokhale, S.M. Goudhari, M.G. Takwale, and S.V. Ghaisas, *High mobility hydrogenated and oxygenated microcrystalline silicon as a photosensitive material in photovoltaic applications*, Appl. Phys. Lett. **60** (1992) 3289-3291.

Hattori, Y., D. Kruangam, T. Toyama, H. Okamoto, and Y. Hamakawa, *High efficiency amorphous heterojunction solar cell employing ECR-CVD produced p-type microcrystalline SiC film*, Tech. Dig. of the International PVSEC-3, Tokyo, Japan (1987) 171.

Kanicki, J. (editor), *Amorphous and microcrystalline semiconductor devices*, Vol. I and II (Artech House, Norwood, MA, 1991).

Lewis, A.J., Jr, G.A.N. Connel, W. Paul, J.R. Pawlik, and R.J. Temkin, in *Tetrahedrally Bonded Amorphous Semiconductors*, eds. M.H. Brodsky, S. Kirkpatrick, and D. Weaire, AIP Conf. Proc. **20** (1974) 27.

Matsuda, A., S. Yamasaki, K. Nakagawa, H. Okushi, K. Tanaka, S. Iizima, M. Matsumura, and H. Yamamoto, Jpn. J. Appl. Phys. **19** (1980) L305.

Matsuda, A., *Formation kinetics and control of microcrystallite in μc-Si:H from glow discharge plasma*, J. Non-Cryst. Solids **59&60** (1983) 767-774.

Paul, W., A.J. Lewis, G.A.N. Connel and T.D. Moustakas, *Doping, Schottky barrier and p-n junction formation in amorphous germanium and silicon by rf sputtering*, Solid State Commun. **20** (1976) 969-972.

Spear W.E. and P.G. LeComber, *Investigation of the localised state distribution in amorphous Si films*, J. Non-Cryst. Solids **8-10** (1972) 727-738.

Spear W.E. and P.G. LeComber, *Substitutional doping of amorphous silicon*, Solid State Commun. **17** (1975) 1193-1196.

Spear W.E. and P.G. LeComber, *Electronic proporties of substitutionally doped amorphous Si and Ge*, Phil. Mag. **33** (1976) 935-949.

Sterling, H.F. and R.C.G. Swann, *Chemical Vapour deposition promoted by r.f. discharge*, Solid-State Electron. **8** (1965) 653-654.

Triska, A., D. Dennison, and H. Fritzsche, *Hydrogen content in amorphous Ge and Si prepared by r.f. decomposition of GeH_4 and SiH_4*, Bull Am. Phys. Soc. **20** (1975) 392.

Usui S. and M. Kikuchi, *Properties of heavily doped gd-Si with low resistivity*, J. Non-Cryst. Solids **34** (1979) 1.

Vepřek, S. and V. Mareček, *The preparation of thin layers of Ge and Si by chemical hydrogen plasma transport*, Solid St. Electr. **11** (1968) 683.

2 DEPOSITION OF AMORPHOUS AND MICROCRYSTALLINE SILICON

...; this clearly demonstrates that the excitation frequency is an important parameter in plasma processing.

—H. Curtins, N. Wyrsch, A.V. Shah, 2nd February 1987

2.1 PLASMA ENHANCED CHEMICAL VAPOR DEPOSITION

Amorphous and microcrystalline silicon thin films are still commonly produced using the method that first resulted in hydrogen incorporation in the material (Sterling and Swann, 1965). The deposition method is a glow discharge technique, also known as Plasma Enhanced Chemical Vapor Deposition (PECVD). A silicon containing gas, usually silane (SiH_4), is admitted to a vacuum reactor chamber. A gas discharge is then initiated and maintained by an electric field between two parallel plates, using either a dc voltage or a voltage in the radio frequency domain (13.65 - 200 MHz). The pressure is typically 0.1 - 1 mbar,

The authors acknowledge H. Meiling from Utrecht University for discussions related to this chapter.

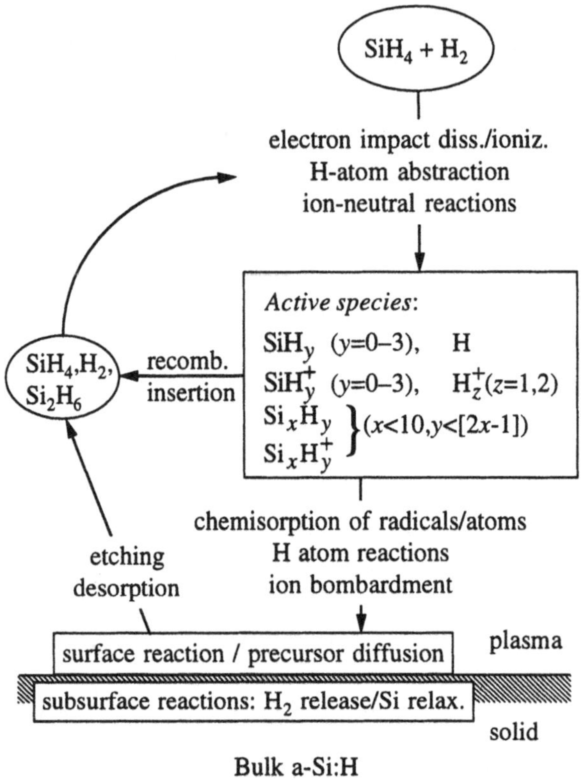

Figure 2.1. Schematic representation of the glow-discharge deposition process (By courtesy of H. Meiling, Utrecht University).

depending on the geometry of the vacuum chamber. The hydrogen bonded to silicon was later recognized as being essential for tighing up the unpaired valence electrons that would otherwise lead to electronic defect states.

The deposition process by an rf discharge can be described as a four step process.

1. The primary reactions in the gas phase are electron-impact excitation, dissociation, and ionization of SiH_4 molecules. The plasma thus consists of

neutral radicals and molecules, positive as well as negative ions, and electrons.

2. Secondary reactions, between molecules and ions or radicals, are very important as they predominantly control the electronic and structural film properties. Reactive neutral species move to the substrate by diffusion, positive ions bombard the growing film, and negative ions are trapped within the sheaths (Perrin, 1995) and may eventually form small particles, or dust.

3. The third step consists of surface reactions, such as hydrogen abstraction, radical diffusion, and chemical bonding.

4. The fourth step is the subsurface release of hydrogen molecules and relaxation of the silicon matrix.

The whole sequence is schematically represented in Fig. 2.1 (Meiling, 1991).

The deposition process is very complicated as the physical and chemical interactions in the plasma and at the growing film surface are dependent on the rf power and frequency, the substrate temperature, the gas pressure and composition, the magnitude and the pattern of the gas flow, the electrode geometry, etc. It is beyond the scope of this book to discuss all effects of these parameters in detail. Recommendable literature on the deposition of hydrogenated amorphous silicon alloys are the textbook authored by Luft and Tsuo, 1993 and the edited book by Bruns et al., 1995.

Amorphous silicon films with good electronic quality are obtained at a low rf power when SiH_3 is the most abundant radical at the surface (Gallagher, 1988). The major fraction of the radicals reaching the surface is not incorporated (Ganguly and Matsuda, 1993), but recombines to, e.g., disilane. The low sticking coefficient and relatively high surface mobility (Perrin et al., 1989) contributes to the formation of a well-configured amorphous silicon network. The surface dangling bonds existing at locations where voids would tend to form locally enhance the sticking of the SiH_3 radical. This mechanism intrinsically inhibits the formation of larger voids so that the material is thus constructed as a dense and homogeneous network.

As under these low discharge power conditions the deposition rate is quite low (< 3 Å/s), much effort has been put in research on compatible methods to increase the deposition rate. A method attracting a lot of attention is the use of a 5-10 times higher discharge frequency, called Very High Frequency CVD (VHFCVD; Curtins et al., 1987), which is discussed in Section 2.1.3.1. Other methods employing PECVD make use of the better dissipation of rf power at a 2-5 times higher pressure, the so called γ regime (Perrin, 1995). It is thought that the negatively charged particles formed in a plasma operated near to or in the γ regime are in fact nanocrystalline and promote a denser material with a lower hydrogen content (Middya et al., 1997). This regime was called γ regime after the classical description of a dc discharge in which many ion-induced secondary electrons are emitted at the cathode. This is accompanied by a significant optical emission from the bulk of the plasma and usually occurs at high pressure and/or power, but it may also occur at relatively high plasma frequency. In contrast, the α regime is characterized by wide sheaths where electrons gain energy from the sheath electric field (Perrin and Schmitt, 1992).

Microcrystalline films can be obtained by PECVD at 13.56 MHz by diluting silane in hydrogen at a ratio of $H_2/SiH_4 > 20$ and relatively high rf power. These films consist of small grains (3 - 50 nm) embedded in an amorphous matrix. They are mainly used for their high conductivity when doped. Concomitantly, they exhibit smaller optical absorption than amorphous silicon due to the indirect band gap component, which makes them yet more useful as doped contact layers in optical devices such as solar cells. The utilization of microcrystalline silicon as the active layer of a solar cell was first reported by Faraji et al., 1992 and this cell had an efficiency of 7.8 % at an illumination intensity of 60 mW/cm^2. Presently, both the VHFCVD method and the HWCVD method have succeeded in preparing highly oriented polycrystalline films with virtually no amorphous tissue. These techniques will be discussed later in this chapter.

2.1.1 Large area deposition

Plasma-enhanced CVD is particularly suitable for producing thin films for large area electronics or so-called giant electronics applications. The flat panel display technology presently requires pinhole-free switching matrices for LCDs on a panel area of 36 cm $\times$ 45 cm, while larger panels of 75 cm $\times$ 95 cm are already envisaged for fourth-generation displays. Other large area applications include 2D image sensing, for offices as well as medical X-ray applications. For photovoltaic applications, 40 cm $\times$ 120 cm is the standard size of NEDO (New Energy and Industrial Technology Development Organization) in Japan. Some manufacturers use a slightly smaller size (40 cm $\times$ 80 cm, Fuji Electric), whereas some produce a range of differently sized modules (United Solar). The now liquidated company APS, Inc., who had to stop production as a result of a patent infringement lawsuit, were technologically successful with their Eureka modules which had an area of 1.2 m^2.

The two basic concerns when going from small to large area are to maintain the material quality and the film uniformity. The ultrahigh vacuum (UHV) practices that are in use in small area laboratory deposition systems to reduce contaminant incorporation cannot be applied to large area systems, due to excessive costs or even lack of availability of UHV parts of sufficiently large size.

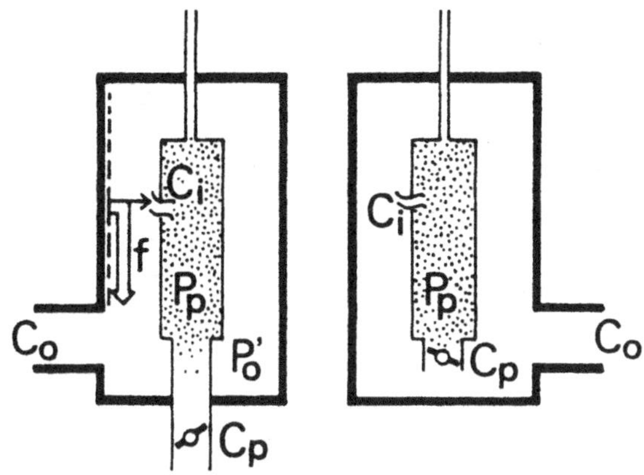

Figure 2.2. Examples of plasma box configurations as proposed by Schmitt et al., 1988. C_0, C_i, and C_p are conductances; P_0 and P_p are the base pressure and the process pressure, respectively. The configuration on the left side uses dual pumping and that on the right side uses internal throttling.

A frequently used approach to simulate UHV conditions in large area systems is the double shell geometry, better known as the plasma-box configuration shown in Fig. 2.2 (Schmitt et al., 1988). The outer shell is pumped via a large opening to a low pressure, while the plasma-box, containing the substrate, is pumped via a throttle valve to the process pressure. The plasma-box is not vacuum tight, but the leakage of impurities from the outer shell to the interior of the plasma box is counterstream and therefore sufficiently small. In systems that do not employ a plasma-box configuration, a similarly low contamination level can be obtained by using a continuous flow arrangement using high purity gases. During the time that no deposition process takes place an inert gas flow is maintained.

Other parameters affecting material quality that do not simply scale with the size of the deposition reactor are the rf power and the gas consumption. Also the flux of ions towards the growing surface and the energy of the ion bombardment are affected when scaling up. The kinetic energy of the ions is mainly dependent on the ratio of the area of the rf driven electrode to that of the grounded counter electrode (including the chamber walls). This ratio is further affected by the presence or absence of plasma confinement shields. When going from small to large area deposition systems, the rf electrode has a relatively large surface by which the above-mentioned area ratio increases to a value close to unity. The ion bombardment energy thereby increases, which may lead to structural damage of the film by ion implantation if this energy becomes ≥ 50 eV.

The depletion of the silane is dependent on the magnitude and pattern of the flow and on the power coupled to the discharge. Depleted silane contains more radicals that are highly reactive, e.g., SiH_2, which contribute to the formation of dust nucleating from negative ions. Depletion of the silane can be overcome in some reactors by merely increasing the flow. In our experience, using a reduced power density and/or hydrogen dilution helps avoiding depletion. The deposition rate is of course limited by this approach.

The second basic concern in large area deposition, the uniformity of the films, has been causing many sleepless nights for many specialists in the field. The most dominant sources for non-uniform deposition are (i) locally inhomogeneous rf impedances due to improper substrate attachment to the grounded electrode (Meiling et al., 1996) or loss of parallelism between rf electrode and substrate, (ii) local changes in the gas composition by depletion away from the point of feedstock injection or by local differences in the residence time by obstruction or recirculation of the gas flow, (iii) local driving voltage variations due to finite rf wavelength effects or due to local extrusions, (iv) local temperature variations.

2.1.2 Production systems

For cost-effective mass production of solar cells, it is required that the product of the amount of time needed per square meter for depositing the multilayer stack and the depreciation and maintenance costs associated with a production machine is small. This means that a low deposition rate can be tolerated if only low capital investment is needed for the production machine and if maintenance is relatively easy. Here, a comparison is made between possible system configurations for a production process, such as *in-line* and *cluster* configurations. Further attention will be paid to different production schemes, such as *single panel*, *batch* and *roll-to-roll* processes. The current production configurations can be classified as follows.

(A) Single chamber system. This system has the advantage that there is no transport of substrates under vacuum conditions. If a load lock chamber is used, a simple form of linear transport is sufficient. The substrates can have a very large size, and can be handled as mounted in a plasma-box. The capital investment is low compared to the other configurations.

A disadvantage is that each run may take a long time since several purge and pump-down steps are needed to obtain well-defined consecutive depositions of differently doped and undoped layers. The throughput can be enhanced by loading many substrates at once. Further, the possibilities for applying buffer layers or graded layers at the interfaces are limited. The geometry of the electrodes and the deposition temperature can not be varied for the various layers. This leads to a trade off between the optimum deposition parameters for the individual layers. Therefore, in a single chamber system the solar cell structure will be less than optimal.

(B) Multichamber system. This system has the advantage of complete control of the level of dopants and other impurities, in basically the same way as in a small area laboratory multichamber system (Madan et al., 1993). In this type of system it is possible to make controlled compositional profiles, with increasing or decreasing concentrations of constituents, without affecting the impurity level of subsequent layers. The deposition temperatures, the internal geometry, and even the discharge frequency can be optimized for each layer individually. The transfer of the manufacturing process from the laboratory to production represents only a small step.

A disadvantage is that the size of the panels may be limited. Nevertheless, panels with a size of 40 cm × 120 cm and 60 cm × 100 cm have been produced with multichamber configurations. Another disadvantage is that the investment cost is quite large. One can distinguish two types of multichamber systems.

(B1) in-line configurations. The throughput of these machines can be very high. The equipment cost however is also very high, because special isolation chambers and vacuum valves are needed between the deposition chambers. Furthermore, in-line machines usually take up quite some floor space. The layer sequence is determined by the physical configuration of the system and can not be adapted easily. This reduces the flexibility in designing the multilayer structure. This configuration offers the possibility to deposit on moving substrates. This leads to extremely good uniformity of the layers over very large areas, an important feature for tandem solar cell structures. As the deposition plasma does not need to be extinguished, an important source of dust and microparticulates is eliminated.

(B2) cluster configuration. This configuration is known from the laboratory, but may also be used for (pilot-)production. A cluster system has all the advantages of a multichamber system. The transition from laboratory experience to production is as small as possible. Modifications in the layer sequence or changes in the number of layers of the solar cell structure (e.g., from single junction to tandem or triple junction) can be made as desired. Apart from this the capital cost is low, because transport- and isolation chambers can be shared by many deposition chambers. If a sufficient number of chambers is connected to the transport chamber, production can be continued even if one of the deposition reactors is down for maintenance. This is in contrast to *in-line* systems which must be shut down completely if one of the chambers needs maintenance. The floor plan of a cluster system is very modest. The flow of reactive gases is only needed when there is a substrate in the deposition chamber, thus cutting down on the base materials usage. Profiled and graded layers are much more easily made than in in-line systems. The presence of a central panel handling system, however, does reduce throughput and is a highly critical component.

All of the above production facilities can be used in a *single panel* mode or in a *batch* mode. In the latter mode a number of substrates is loaded at the same time, a principle often applied in a single chamber system to bring throughput to the required level. Due to the long processing time, single chamber systems can only be cost-effective when they are batch-type systems.

(C) roll-to-roll configuration. A very effective method to improve throughput is the roll-to-roll method. This method is used by United Solar and its business partners. Fuji Electric uses the "stepping" roll-to-roll method, which does not utilize continuous motion. In continuous motion, only one dimensional uniformity is required. The transport mechanism can be cheaper than in most panel-by-panel in-line systems, as there are no driving mechanisms in the plasma regions. The thin web, e.g., stainless steel, is light weight and can therefore be heated and cooled quickly. The various amorphous silicon layers and alloys can be deposited in a single pass. Dual- or triple-junction cells merely require a larger number of deposition zones.

As the substrate is continuous, all deposition zones are by definition connected to each other. To prevent cross contamination, special gates have to be incorporated which dynamically isolate the adjacent zones. The "gas gates" developed at Energy Conversion Devices, Inc. (ECD) utilize laminar gas flow through a constant geometrical cross section conduit in a direction opposite to the diffusion gradient of the dopant gas.

2.1.3 Compatible PECVD techniques

2.1.3.1 Very High Frequency CVD. A PECVD technique that has attracted considerable attention is the Very High Frequency CVD technique (Curtins et al., 1987a, Curtins et al., 1987b), in which the plasma excitation frequency is increased from the conventional 13.56 MHz into the VHF range. An attractive feature of this technique is the 5 - 10 times higher deposition rate for amorphous silicon films, which is commonly thought to contribute to a lower production cost of a-Si:H-based solar cells. The maximum achievable deposition rate is system dependent, where proper impedance matching of the rf power to the glow discharge is crucial. In contrast to the first observations by Curtins et al., 1987b, usually a monotonic increase of the deposition rate with the excitation frequency is found, while the material properties are more dependent of other deposition parameters such as the pressure.

The explanation for the increased deposition rate is not straightforward. From decomposition measurements in a silane plasma using mass spectromety it was shown that the increased deposition rate can not be simply the result of enhanced radical production (Bezemer and Van Sark, 1994). Therefore it is thought that the increased ionization rate due to the relatively high concentration of electrons in the high energy tail of the distribution plays a dominant role (Chatham and Bhat, 1989, Heintze et al., 1993). Vepřek et al., 1989 showed that an increased flux of low energy ions enhances hydrogen abstraction from the surface (see Fig. 2.1), which promotes the bonding of radicals to the surface. Indeed, Chatham and Bhat, 1989 already deduced from measurements of the applied rf voltage and the dc self bias voltage, that the plasma potential, and thus the mean ion energy, decreases while the increased ionization rate with increasing frequency results in a larger ion flux. At the same time, the low-energy, high-flux ion bombardment is believed to enhance the surface mobility of adsorbed species, so that the properties of the materials remain device-quality also at a high deposition rate. Silicon-ion bombardment energies higher than 16 eV are believed to give rise to defect formation (Vepřek et al., 1989). The value of 16 eV might however be underestimating the critical value for the case of a-Si:H deposition, as defects due to ion impact still have a chance to recover due to subsurface silicon matrix relaxation (Fig. 2.1).

Dutta et al., 1992 estimated the peak ion energy for silicon ions as a function of the exitation frequency, see Fig. 2.3, and showed that an excitation frequency larger than 50 MHz prevents the occurrence of Si ions with energies higher than the above-mentioned threshold. Hamers et al., 1998 showed that the ion bombardment could be quantified by introducing the kinetic ion energy per *deposited* silicon atom. At a deposition temperature of 250 °C, values in excess of 5 eV per deposited atom are necessary for obtaining a dense amorphous silicon network with little microstructure (the microstructure parameter is defined in Section 3.1.1).

When using high rf frequencies, other deposition parameters, such as process pressure and electrode geometry need to be re-optimized. It has been found that the maximum of the SiH* emission peak at 414 nm shifts to a lower pressure at increasing frequency (Chatham and Bhat, 1989). Therefore, to take full advantage of the enhanced excitation and dissociation rate it is advisable to

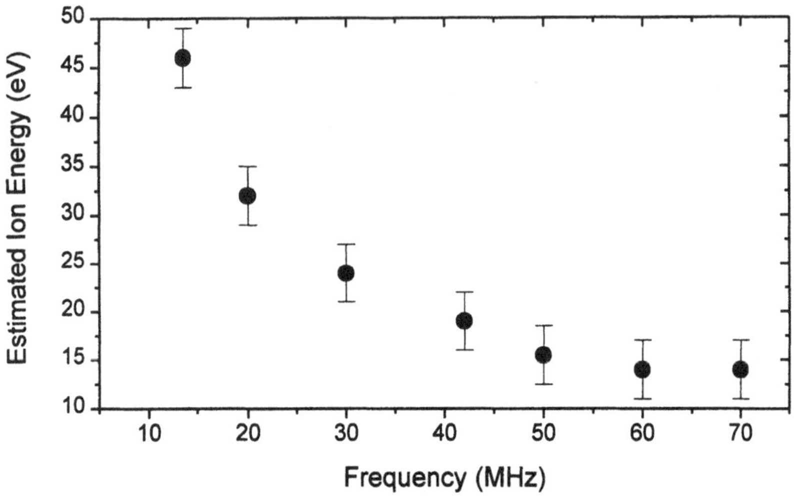

Figure 2.3. Estimated peak Si-ion energies for SiH_4 plasmas as a function of excitation frequency (after Dutta et al., 1992).

reduce the operating pressure when using high rf discharge frequencies. As the radical surface mobility is maintained due to the high ion flux at these frequencies, the VHFCVD technique allows the use of a reduced deposition temperature which can be quite helpful in device applications.

Commonly, as shown in Fig. 2.4 (Schropp et al., 1989), the initial conversion efficiency of p-i-n solar cels is suppressed at high PECVD rates due to inferior electronic transport characteristics. In addition, the light-induced degradation of cells incorporating a high deposition rate intrinsic layer is usually much more severe than that of those made with a very low deposition rate intrinsic layer (Xi et al., 1987). Amorphous p-i-n solar cells which have an intrinsic layer deposited at a rate of 18 Å/s by VHFCVD, however, showed an initial efficiency of 9.7 % (Chatham et al., 1989b). The amount of light-induced degradation of these cells is similar to that of cells with a standard intrinsic layer deposited at 2 - 3 Å/s. More recently, Jones et al., 1998 found that there is a clear advantage to using the VHF technique up to deposition rates of 10 Å/s. In this range efficiencies and amounts of degradation for n-i-p solar cells were found to be similar to those of cells made at 13.56 MHz and a deposition rate of 1 Å/s.

If the silane feedstock gas is diluted with hydrogen, microcrystalline silicon is readily obtained by VHFCVD as first demonstrated by Oda et al., 1988. These microcrystalline layers are interesting for use as highly doped p- and

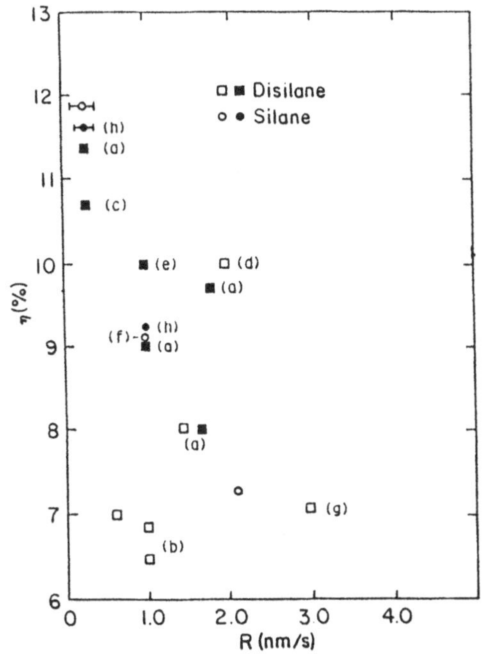

Figure 2.4. The initial efficiency of single-junction a-Si:H solar cells plotted as a function of intrinsic layer deposition rate. The open symbols refer to cell areas of less then 0.1 cm^2, the filled symbols refer to cell areas of at least 0.1 cm^2. (Reprinted from Schropp et al., 1989, with permission from Elsevier Science. For sources of data, see the same literature reference).

n-type contact layers in solar cells. These contact layers are more transparent due to their indirect gap component and they can be more efficiently doped. Recently, microcrystalline *intrinsic* silicon made by VHFCVD has been demonstrated to be suitable for application as the i-layer in p-i-n type solar cells on glass (Faraji et al., 1992, Meier et al., 1994). This possibility is due to the fact that VHFCVD microcrystalline films have needle-like oriented crystallites in the direction of the carrier transport with a typical diameter of 15 nm, while their boundaries are hydrogen-passivated due to the low-temperature deposition conditions. A stable microcrystalline p-i-n solar cell was reported by Meier et al., 1995, with an efficiency of 7.7 % . The microcrystalline intrinsic absorber has a band gap of 1.0 eV (Beck et al., 1996) and a thickness of 3.6 μm. The low band gap derived from extrapolated absorption spectra was later explained by enhanced scattering (see Section 3.1.2). The films have high tensile stress and are further characterized by a hydrogen content of about 5 at.-%. Recently, the Neuchâtel group improved the microcrystalline p-i-n cell to 8.5 % (Meier et al., 1998, area 0.25 cm^2), however no stability data were reported for this cell thus far. At Kaneka Corp., Japan, a thin film polycrystalline n-i-p cell has recently been developed with an aperture area efficiency of 10.1 % (Yamamoto et al., 1998). It is not known which rf frequency was used during plasma deposition of this cell. A drawback of the microcrystalline intrinsic layer is the low deposition rate of 1 - 2 Å/s, especially since this layer has to be made rather thick compared to the intrinsic layer of an amorphous solar cell. Introducing argon to the discharge increases the deposition rate as the Ar atoms achieve an excited metastable state in the plasma which contributes to the dissociation of silane molecules (Meiling et al., 1997a). However, Ar dilution has not resulted in equivalent solar cell performance at elevated deposition rates so far (Keppner et al., 1996).

The application of VHFCVD in the mass production of large area solar panels with either amorphous or microcrystalline components might be limited due to the finite wavelength λ associated with the rf frequency. If an rf electrode has a single feed point on one side a standing wave may develop if the electrode has a length of $\lambda/4$ or smaller. At 13.56 MHz, λ is 22 m, and does not impose practical limitations to the size of the area to be coated. At moderately elevated frequencies the finite wavelength effect might not have a large influence on the uniformity of the deposited layers. However, at frequencies > 60 MHz, the effect of standing waves might become noticeable in large area (≈ 1 m^2) deposition. Here, multipoint feeding of the rf power with equal amplitude and phase improves the homogeneity of the rf potentials.

2.1.3.2 Hydrogen diluted PECVD. Since 1996 there is renewed attention for hydrogen dilution of the feedstock gas as a method to influence the properties of the deposited layers. In particular, hydrogen dilution of silane has appeared to be beneficial to the structural and electronic properties of wide band gap a-Si:H (Xu et al., 1993) as well as narrow band gap a-SiGe:H (Xu et al., 1996), for multijunction cell components, and it improves the stability of standard band gap single junction cells (Okamoto et al., 1996) as well as wide band gap single junction cells (Rech et al., 1995). The effect of hydrogen dilution of SiH_4 on the materials properties of a-Si:H was in fact already described much earlier by Chaudhuri et al., 1984 and Shirafuji et al., 1985, but these early results were sometimes conflicting. The explanation put forward today more properly takes the degree of dilution into consideration. At moderate dilution, up to a ratio $r = H_2/SiH_4$ of 2, the total hydrogen content decreases, the microstructure parameter R^* (see Section 3.1.1 for definition) decreases to below 0.05, the optical band gap decreases, and the photoresponse improves (Meiling et al., 1990). The same effects were elucidated by Okamoto et al., 1996 and were extended to higher dilution ratios r up to 20. The latter publication reconciles the divergent results in earlier work. In the extended range ($r > 20$) the hydrogen content again increases, R^* remains lower than 0.05, the photoresponse further improves, and the band gap increases again. The dilution regime between $5 < r < 20$ yields wide band gap amorphous layers with better optoelectronic quality than layers obtained from pure SiH_4 or SiH_4-CH_4 mixtures. At still higher dilutions, inhomogeneous microcrystalline layers are obtained.

It is not exactly clear, however, which is the main mechanism responsible for the improved materials properties upon dilution with hydrogen. Possible mechanisms include

- the atomic hydrogen produced in the hydrogen plasma etches strained bonds that are energetically less stable, resulting in a dense network (Tsai et al., 1988),
- hydrogen scavenges deleterious radicals with a short lifetime (e.g., SiH_2) that would otherwise lead to pronounced microstucture (Ganguly and Matsuda, 1996),
- excess hydrogen saturates the growing surface and increases the diffusion length of the growth precursors (e.g., SiH_3) (surface reaction model; Matsuda, 1983),
- hydrogen atoms diffuse into the subsurface growth zone where they help reconstruct the silicon network into a denser network (growth zone model; Shibata et al., 1987, Shirai et al., 1994),
- hydrogen ions may play an additional role in the subsurface growth zone as their kinetic energy may enhance the network reconstruction. On the other hand they may reduce the hydrogen content due to ion-induced desorption.

Also a-SiGe:H has greatly enhanced structural properties when hydrogen dilution is employed during deposition. The beneficial effects of hydrogen dilution were in an early stage attributed to the prevention of Ge clustering and dihydride bond formation (Yang et al., 1991). High hydrogen dilution has been shown to be particularly useful for the deposition of good quality low band gap a-SiGe:H (1.32 eV; cubic gap, see Section 3.1.1 for definition) at a low temperature (180°C) (Kinoshita et al., 1997).

A drawback of hydrogen dilution is the decreased deposition rate, but this is unavoidable as the enhanced network reconstruction either by etching or by growth zone reconstruction does take time. Hydrogen dilution also avoids depletion of the silane. This is particularly helpful in large area deposition, where heavily depleted silane would otherwise start forming dust or where depletion would lead to nonuniform thickness of the deposited films.

Still conflicting results are sometimes found in different laboratories. The geometry of the deposition system determines to a large extent whether hydrogen dilution is beneficial or not. In systems where the interelectrode distance d is relatively large, the product of process pressure p and this distance d, the pd product, would become large. A large pd product makes the plasma operate in the γ regime which stimulates dust formation and yields non-uniform films. In these systems hydrogen dilution is beneficial. If the interelectrode distance is relatively short, hydrogen dilution is found to have no effect (Daey Ouwens et al., 1993).

2.2 ALTERNATIVE CVD APPROACHES

2.2.1 Hot-wire deposition

The method of catalytic decomposition of silane or silane/hydrogen mixtures at a resistively heated filament (nowadays generally called Hot Wire Chemical Vapor Deposition; HWCVD) has been utilized to produce thin films of polycrystalline and amorphous silicon at relatively low temperatures on a variety of substrates.

The method of Hot Wire CVD was first introduced and patented in 1979 as thermal CVD by Wiesmann et al., 1979, who made use of a tungsten or carbon foil that was heated to a temperature of 1400 - 1600 °C. At such temperatures thermal decomposition of silane into a gaseous mixture of hydrogen and silicon atoms was thought to take place. The method was further developed by Matsumura, 1986 as catalytic chemical vapor deposition (CTLCVD), using a heated wire instead of a foil. The technique was further studied by Doyle et al., 1988, Matsumura, 1988, and Matsumura, 1989 showing a high deposition rate as the prominent feature. Typical for the work by Matsumura was the relatively low filament temperature (1200 - 1350 °C) in combination with a higher pressure (> 0.1 mbar) and a larger filament surface area (> 5 cm^2) than is commonly used nowadays.

Until 1991, the catalytic or thermal CVD processes did not consistently yield amorphous silicon films with a quality equal to that of device-quality films obtained by PECVD. The main reason for this was that the early research in this field was carried out in an unsuitable parameter regime. In particular, the combination of the small filament-substrate distance and large filament surface area was leading to a too large flux of Si atoms at the growing surface. Atomic Si has a high sticking coefficient, which leads to poor quality films if it reaches the film surface in a large amount. However, reasonably good films could still be obtained under these circumstances if the wire temperature was chosen below

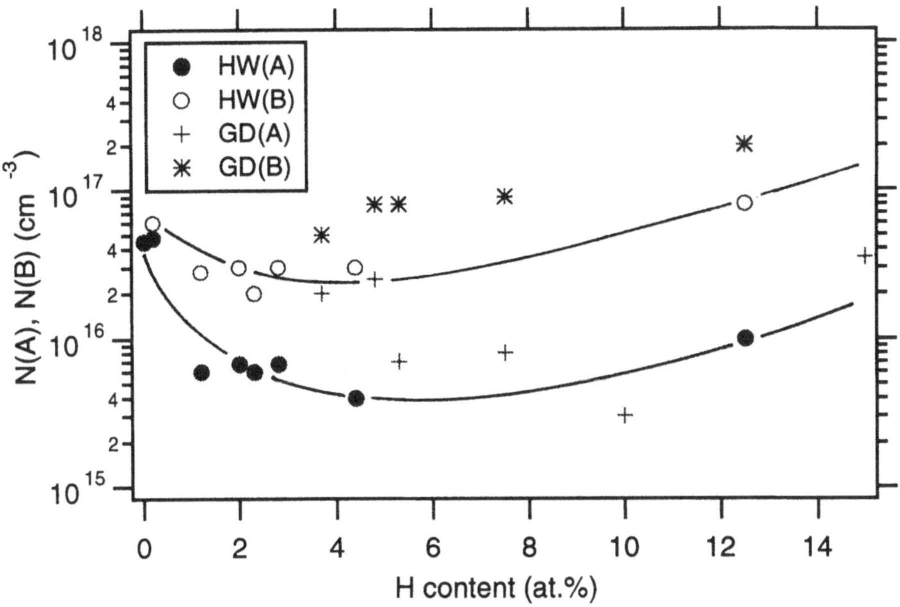

Figure 2.5. Density of deep states N_d versus hydrogen concentration C_H in the fully light-soaked state (By courtesy of H. Mahan, National Renewable Energy Laboratory).

1500 °C, as the wire may scavenge a large fraction of the atomic Si radicals at this temperature.

Inspired by the work of Gallagher's group, Mahan et al., 1991a, Mahan et al., 1991b, and Crandall et al., 1992 demonstrated for the first time the possibility to produce device-quality a-Si:H with a hydrogen concentration below 1 at.-% (see Fig. 2.5), which triggered new interest in the deposition method which they renamed to Hot Wire CVD. The principle of the success of HWCVD in obtaining these exceptionally good films is that the feedstock gas, such as silane, is very effectively decomposed into atomic fragments at the surface of the filament if this is kept at a temperature significantly higher than 1500 °C,

$$SiH_4 \rightarrow Si + 4H. \tag{2.1}$$

In combination with a low pressure this enables a high deposition rate without gas-phase nucleation of particles.

Zedlitz et al., 1993, and Heintze et al., 1996 confirmed that silane dissociation is indeed induced by a surface catalytic effect, as claimed by Matsumura, 1988, and Matsumura, 1989. In contrast to the conventional PECVD technique, no

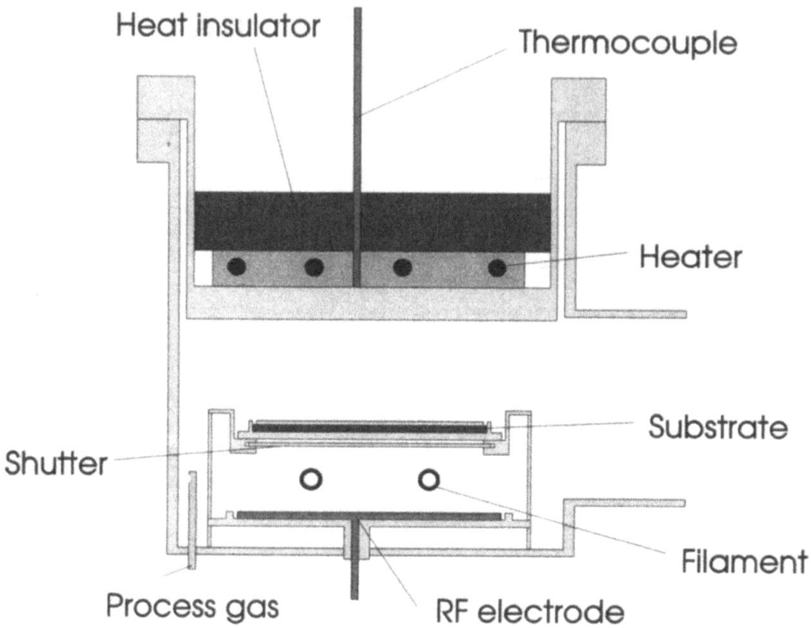

Figure 2.6. Schematic cross section of an HWCVD reactor. The same chamber can be used for PECVD.

ions are created: though the hot filament emits a considerable electron current, the energy of these electrons is generally too low to cause impact ionization.

A schematic cross section of an HWCVD set up is shown in Fig. 2.6. The substrate is placed a few centimeters from the filament. The wire used as the hot filament is made of tungsten or tantalum and has a thickness between 0.25 and 0.5 mm. The wire is resistively heated by a dc or an ac current. The wire temperature is preferably kept between 1750 °C and 1950 °C. Lower temperatures lead to tungsten silicide formation on the wires, thus reducing the catalytic action of the hot filament. Higher temperatures lead to evaporation of the tungsten wire itself and subsequent detectable tungsten incorporation in the film. By employing a proper multi-wire arrangement it is possible to obtain good thickness uniformity over a large area. The wires are acting as true line sources (Feenstra et al., 1997).

The gas flow is chosen as to prevent depletion of the silane. The gas pressure and gas source composition (e.g., silane diluted in hydrogen) have a large influence on the nature of the precursors and the structure of the deposited film. Good quality a-Si:H films are generally obtained from pure silane, although a source gas of silane diluted to 1 % in He also leads to good material properties (Molenbroek et al., 1996). Dusane et al., 1993 demonstrated the possibility of obtaining microcrystalline films from pure silane at a wire temperature of 1400 °C. Brogueira et al., 1996, and Heintze et al., 1996 described the transition from amorphous to microcrystalline silicon deposition upon dilution of silane in hydrogen.

For a good quality amorphous silicon film, a moderate amount of secondary gas phase reactions is desirable. The main secondary reactions are believed to be the insertion reaction (Molenbroek, 1995a),

$$\mathrm{Si} + \mathrm{SiH_4} \rightarrow \mathrm{Si_2H_4^*}, \tag{2.2}$$

and the hydrogen abstraction reaction,

$$\mathrm{H} + \mathrm{SiH_4} \rightarrow \mathrm{SiH_3} + \mathrm{H_2}. \tag{2.3}$$

If the pressure is too high or the substrate-filament distance is too large, film properties will be inferior due to significant gas phase polymerization. If the opposite deposition conditions are in effect, the film properties will suffer from a high flux of Si atoms arriving at the surface, leading to a large concentration of voids and a pronounced microstructure. The concentration of atomic hydrogen in HWCVD plays a role in creating the right radicals for deposition (SiH_3) as well as in balancing the hydrogenation and etching of the growing surface. It has recently been shown that the atomic hydrogen flux effectively rehydrogenates the film *during* growth to a depth of $\approx$ 200 nm (Feenstra, 1998). Controlling this flux thus is an important prerequisite for producing amorphous silicon thin films with a low hydrogen content.

Good quality poly-Si:H films are obtained when the silane is diluted in hydrogen at a higher pressure. The hot filament decomposes all hydrogen molecules into reactive atoms that effectively etch silicon from disordered or strained bonding sites, leading to a transition from an amorphous network to a crystalline network. The reason why HWCVD is so much more successful in producing polycrystalline films without any amorphous tissue than PECVD using highly diluted silane is that the filament is a much more effective source for *atomic* hydrogen than a glow discharge (Jansen et al., 1989). The surface etching activity is therefore enhanced, but at the same time, more gas phase hydrogen abstraction reactions (Reaction 2.3) take place that help maintain a reasonable growth rate (> 5 Å/s). In contrast, in PECVD the bombardment of H^+ ions inhibits the growth of larger crystallites as was demonstrated by Kondo et al., 1996. Only if the energy of the H^+ ions is reduced, e.g., in a triode reactor or by increasing the excitation frequency, the crystallinity of microcrystalline silicon can be enhanced. In HWCVD, upon accurate tuning of the deposition regime, larger grains are obtained and the amorphous phase in these microcrystalline materials can be completely eliminated, i.e., coalescence is almost perfect. In these materials there are hardly any grain boundary regions, i.e., the grain boundaries are reconstructed and only a very small hydrogen concentration (< 0.5 %) is required to passivate them. These materials are defined as polycrystalline materials, rather than microcrystalline (see Table 1.1). Especially for materials at the extremes of the amorphous - crystalline scale (i.e. purely amorphous or purely polycrystalline), superior properties can be obtained by HWCVD as compared to other deposition methods. Typical deposition conditions are listed both for amorphous and polycrystalline films in Table 2.1. Note that for the extreme cases of completely amorphous and completely polycrystalline films, no changes have to be made in the hot-wire assembly or the internal geometry of the chamber; the respective films can be obtained merely by altering the feed gas composition, the pressure, the wire temperature, and the substrate temperature.

Table 2.1. Typical deposition conditions used for hot-wire deposition of amorphous and polycrystalline intrinsic silicon films.

Parameter	*a-Si:H*	*poly-Si:H*
Wire temperature (°C)	1900	1800
Substrate temperature (°C)	430	480
SiH_4 flow (sccm)	90	10
H_2 flow (sccm)	0	100
Process pressure (mbar)	0.02	0.1

Whereas a-Si:H made by PECVD requires a hydrogen content of 8 - 12 at.-% in order to obtain a low density of defect states, long carrier lifetimes, sharp Urbach edges and a low void density, HWCVD has been demonstrated to yield the same quality of material with only 1 - 3 at.-% hydrogen. The Urbach energy remains in the 50 - 60 meV range, even at a hydrogen concentration of 0.2 at.-% (Mahan et al., 1991a). These materials can be expected to show less light-induced degradation according to models that link the excess hydrogen in microstructural defects to light-induced metastable defect creation (Zafar and Schiff, 1989). Indeed, the metastable state for HWCVD a-Si:H has superior electronic properties to that of PECVD a-Si:H (Mahan and Vaněček, 1991, Papadopoulos et al., 1993). Moreover, with a deposition rate in excess of 10 Å/s, HWCVD produces device-quality layers at a rate that is more than an order of magnitude higher than for conventional 13.56 MHz PECVD. Deposition rates of 50 Å/s (Molenbroek et al., 1997) for high electronic quality a-Si:H have been demonstrated with HWCVD, and even up to 70 Å/s (Heintze et al., 1996), though these films possessed slightly lower quality. The first attempts to incorporate HWCVD a-Si:H in a solar cell were undertaken independently by Papadopoulos et al., 1993 and by Nelson et al., 1994. Recently, a 9.7 % p-i-n cell (with an area of 0.082 cm^2) was reported by Bauer et al., 1997. The HWCVD intrinsic layer of this cell was made at such a low temperature that the light-induced degradation appeared similar to PECVD material. It thus appears to be useful to take full advantage of the capability of the HWCVD technique to combine good electronic quality with a low H content in the film which is achieved at a relatively high substrate temperature of $\geq$ 320°C. A 9.8 % n-i-p cell has been reported by Mahan et al., 1998. The i-layer was deposited at a high rate of 16 Å/s and its hydrogen content was roughly 4 - 5 %. No results on light soaking tests are yet available. Recent work to incorporate HWCVD a-Si:H in thin film transistor structures has led to the first demonstration of stable device operation (Meiling and Schropp, 1997b).

Also poly-Si:H thin films made by HWCVD have superior properties compared to films made by other techniques. Only recently, in 1994, the first high quality poly-Si:H layers have been prepared by HWCVD (Cifre et al., 1994, Middya et al., 1995), although these layers were not yet incorporated in devices. These films showed a high crystalline volume fraction and a low density of defect states. The first application of poly-Si:H in solar cells was reported in 1996 (Rath et al., 1996). Similar to textured plasma-deposited cells (Yamamoto et al., 1998), these cells showed the capability to generate currents in excess of 23 mA/cm^2 (Rath et al., 1998). These poly-Si:H layers could also be applied in stable thin film transistors (TFTs) (Rath et al., 1997, Meiling et al., 1998). Matsumura et al., 1997 demonstrated good polycrystalline characteristics with field-effect mobilities of 70 $cm^2V^{-1}s^{-1}$. The most promising techniques for the preparation of microcrystalline or polycrystalline silicon films other than the HWCVD technique or the VHFCVD technique are the remote microwave plasma technique (Ishihara et al., 1994) or the layer-by-layer technique using PECVD at 13.56 MHz (Ishihara et al., 1993) or similar methods known as *chemical annealing* (Shirai et al., 1991). The layer-by-layer technique has been successful in terms of microcrystalline material quality, even at a very low hydrogen concentration. However, also in this technique the effective film fabrication rate ($\approx$ 1 Å/s) is lower than in HWCVD. Typical materials properties of purely amorphous and purely polycrystalline intrinsic silicon films deposited at a high deposition rate utilizing HWCVD are discussed in Schropp et al., 1997.

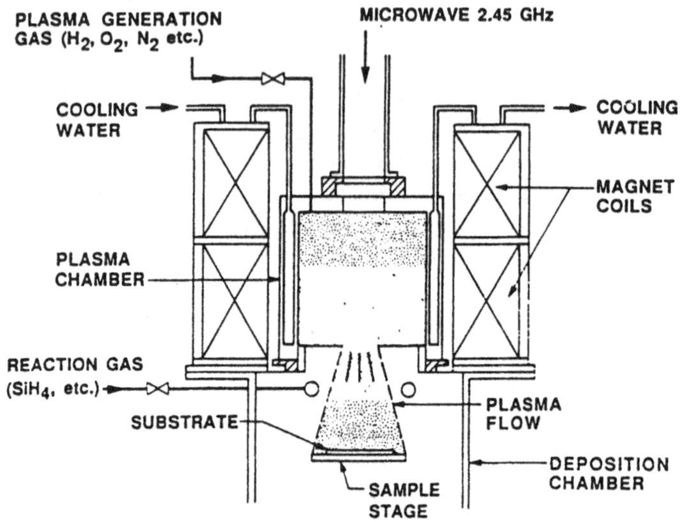

Figure 2.7. Schematic diagram of an ECR microwave CVD reactor (Reprinted from Shing, 1989, with permission from Elsevier Science).

2.2.2 ECR, microwave deposition, plasma-beam deposition

ECR-CVD. The Electron Cyclotron Resonance CVD (ECR-CVD) method (Ichikawa et al., 1987) is in fact a remote-plasma CVD method. Fig. 2.7 shows a schematic diagram. A plasma is generated in a carrier gas such as Ar, He of H_2 by microwave excitation (usually 2.45 GHz) in a remote chamber. Due to an externally applied magnetic field the electrons achieve cyclotron motion and by tuning the field strength B the frequency of the electrons $\omega = \mathrm{e}B/\mathrm{m}$ is brought at resonance with the microwave frequency. Thus the remote plasma chamber acts as a resonance cavity and hence the method is called Electron Cyclotron Resonance CVD. The microwave power is very efficiently transferred to the plasma. The plasma is characterized by a high density of high energy electrons, which leads to a high dissociation rate, while the gas temperature is low. The plasma can be sustained at pressures as low as 1 mTorr.

The feedstock silane is introduced at a location near the substrate. As the magnetic field diverges into the deposition chamber, an electic field is generated in the plasma column, and ionized radicals (SiH_n^+) from the plasma are driven to the substrate where they bombard the growing surface. As discussed in the section on VHFCVD (Section 2.1.3.1), a large ion flux with a sufficiently low energy enhances the deposition rate while material quality is preserved. The large ion flux allows the use of a low substrate temperature. The ion energy can be controlled by the gas pressure. A deposition rate of 25 Å/s for near device-quality a-Si:H has been achieved (Shing, 1989, Watanabe et al., 1987). Due to the significant ion energy transfer to the substrate, ECR-CVD is particularly suitable for a-SiGe:H deposition (Shing, 1989) and for highly conductive microcrystalline doped layer deposition (Hattori et al., 1987). Dalal et al., 1994 proposed that more stable amorphous silicon materials could be produced using H_2-dilution ECR-CVD. Recently, this technique for intrinsic a-Si:H has been applied to n-i-p solar cells (Dalal et al., 1997).

Microwave CVD. The method referred to as microwave CVD method can be either an afterglow CVD technique or a technique where the substrate is in direct contact with the plasma.

The *remote* microwave CVD method is much like ECR-CVD, but a higher magnetic field is used so that the electrons are not at resonance. This method is more suitable for large area deposition as high ion densities can be achieved in a diverging plasma. Again, a low process pressure is used and high deposition rates can be achieved. An inherent drawback with respect to ECR-CVD is the decreased density of the plasma at the growing surface and the reduced gas utilization efficiency.

In contrast, the MPCVD (Microwave Plasma CVD) technique, in which the substrate is in contact with the microwave plasma, has in addition to large area capability the advantage of high deposition rates (> 100 Å/s). The method is specifically interesting for large area a-SiGe:H deposition at a high deposition rate and has been developed (by Canon, Japan) for roll-to-roll coating of conductive web substrates.

Plasma Beam Deposition. The plasma beam deposition (PBD) method can also be considered a remote CVD technique. The method was originally in use for the high rate synthesis of diamond from methane (Kurihara et al., 1988) and has now been adapted for the deposition of amorphous silicon at a high deposition rate (Severens et al., 1996). Fig. 2.8 is a diagram of plasma beam reactor. A thermal plasma is produced in argon or in an argon/hydrogen mixture by a dc arc or a cascaded arc discharge. The plasma beam expands into a reactor chamber and is directed at the substrate. The feedstock silane is injected just after the torch nozzle and is dissociated by a charge-exchange reaction with the ionized argon atoms,

$$SiH_4 + Ar^+ \rightarrow SiH_3^+ + Ar + H. \tag{2.4}$$

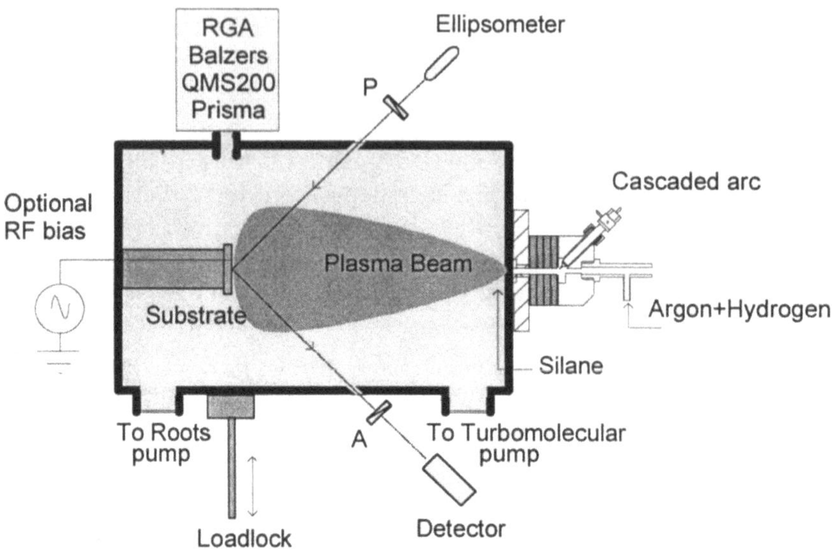

Figure 2.8. Schematic diagram of a cascaded arc plasma beam CVD reactor. (By courtesy of M.C.M. van de Sanden, Eindhoven University of Technology).

Subsequently, the main dissociation mechanisme after expansion is the hydrogen abstraction reaction (Reaction 2.3). The method is interesting because of the following features: (i) the directed plasma beam results in a high radical flux at the substrate which leads to high deposition rate, (ii) the substrate is not part of a capacitive configuration as in PECVD and therefore ions are not accelerated to the growing surface, (iii) the flux of atomic hydrogen is high, which aids hydrogen abstraction reactions at the growing surface.

Thus in many respects the PBD method resembles the HWCVD method, as the gas dissociation is more complete, which generates a high density of radicals along with an abundance of hydrogen atoms. Further, both methods do not feature ion bombardment. Like ECR-CVD, the disadvantage of PBD is that it can not easily be scaled up to large-area deposition as the beam has a pronounced profile and a larger diameter configuration could not be made in a compact way. The increased equipment cost for large area systems might defeat cost advantages due to the high deposition rate. However, PBD does lead to interesting new insight in the deposition mechanisms for amorphous silicon (Kessels et al., 1998).

Table 2.2. Characteristic quantities of various deposition techniques (laboratory systems) for the deposition of device-quality hydrogenated amorphous silicon.

Parameter	*PECVD (13.56 MHz)*	*VHFCVD (60-80 MHz)*	*ECRCVD*	*HWCVD*	*PBD (downstream)*
Electron density n_e (cm^{-3})	10^9	$5-10\times10^9$	10^{11}	0-10^{11}	10^{11}
Ionization ratio	10^{-5}	10^{-4}	10^{-2}	0	10^{-3}
Sheath voltage (V)	40-50	5-15	0.2	0	2
Deposition rate (Å/s)	2	20	25	20	100
Process pressure (mbar)	0.5	0.2	10^{-3}	10^{-2}	0.2
Ar gas flow (sccm)	-	-	-	-	3000
H_2 gas flow (sccm)	-	-	20	-	300
SiH_4 gas flow (sccm)	20	20	20	60	600
Decomposition power consumption (W)	4	4	300	240	5000
Typical sample size (cm^2)	100	100	25	60	25
Typical deposition temp. (°C)	250	200	200	350	450

References

Bauer, S., W. Herbst, B. Schröder, and H. Oechsner, *a-Si:H solar cells using the Hot-Wire technique - how to exceed efficiencies of 10 %*, Proc. 26th IEEE PVSC, Anaheim, CA, (1997) 719-722.

Beck, N., J. Meier, J. Fric, Z. Remes, A. Poruba, R. Flückiger, J. Pohl, A. Shah, and M. Vaněček, *Enhanced optical absorption in microcrystalline silicon*, J. Non-Cryst. Solids **198-200** (1996) 903-906.

Bezemer, J., and W.G.J.H.M. van Sark, *Structure of amorphous silicon, deposited by VHF plasmas*, in: *Electronic, Opto-electronic and Magnetic Thin Films*, Proceedings of the 8th International School on Condensed Matter Physics (ISCMP), Varna, Bulgaria, 1994, Edited by J.M. Marshall, N. Kirov, and A. Vavrek (John Wiley & Sons, New York, 1995) 219-226.

Brogueira, P., J.P. Conde, S. Arekat, and V. Chu, *Amorphous and microcrystalline silicon films deposited by hot-wire chemical vapor deposition at filament temperatures between 1500 and 1900 °C*, J. Appl. Phys. **79** (1996) 8748-8760.

Burno, G., P. Capezzuto, and A. Madan, *Plasma deposition of amorphous silicon-based materials* (Academic Press, Inc., 1995).

Chatham, H., and P.K. Bhat, *High deposition rate intrinsic amorphous silicon materials and p-i-n devices from disilane*, in: "Oct. 1989 SERI Subcontractors Review Meeting", Golden, Co." (SERI/STR-211-3562, 1989).

Chatham, H., P. Bhat, A. Benson, and C. Matovich, *High-efficiency amorphous silicon p-i-n solar cells deposited from disilane at rates up to 2 nm/s using VHF discharges*, J. Non-Cryst. Solids **115** (1989b) 201-203.

Chaudhuri, P., S. Ray, and A.K. Barua, *The effect of mixing hydrogen with silane on the electronic and optical properties of a-Si:H thin films*, Thin Solid Films **113** (1984) 261-270.

Cifre, J., J. Bertomeu, J. Puigdollers, M.C. Polo, J. Andreu, and A. Lloret, *Polycrystalline silicon films obtained by hot-wire chemical vapour deposition*, Appl. Phys. A **59** (1994) 645-651.

Crandall, R.S., A.H. Mahan, B.P. Nelson, M. Vaněček, and I. Balberg, *Properties of hydrogenated amorphous silicon produced at high temperature*, AIP Conf. Proc. **268** (1992) 81-87.

Curtins, H., N. Wyrsch, and A.V. Shah, *High-rate deposition of amorphous hydrogenated silicon: effect of plasma excitation frequency*, Electronics Lett. **23** (1987) 228-230.

Curtins, H., N. Wyrsch, M. Favre, and A.V. Shah, *Influence of Plasma Excitation Frequency for a-Si:H thin film deposition*, Plasma Chem. and Plasma Processing **7** (1987a) 267-273.

Curtins, H., N. Wyrsch, M. Favra, K. Prasad, M. Brechet, and A.V. Shah, *Influence of plasma excitation frequency on deposition rate and on film properties for hydrogenated amorphous silicon*, in: Amorphous Silicon Semiconductors - Pure and Hydrogenated, edited by A. Madan, M. Thompson, D. Adler, and Y. Hamakawa, Materials Research Society Symp. Proc. **95** (1987b) 249-253.

Daey Ouwens, J., R.E.I. Schropp, C.H.M. van der Werf, M.B. von der Linden, C.H.M. Marée, W.F. van der Weg, P. Rava, F. Demichelis, C.F. Pirri, and E. Tresso, *Effect of electrode spacing and hydrogen dilution on a-SiC:H and a-Si:H layers*, in: Amorphous Silicon Technology - 1993, edited by E.A. Schiff, M.J. Thompson, A. Madan, K. Tanaka, P.G. LeComber, Materials Research Society Symp. Proc. **297** (1993) 61-66.

Dalal, V.L., E.X. Ping, S. Kaushal, M. Bhan, and M. Leonard, *Growth of high-quality amorphous silicon films with significantly improved stability*, Appl. Phys. Lett. **64** (1994) 1862-1864.

Dalal, V.L., T. Maxson, R. Girvan, and S. Haroon, *Stability of single and tandem junction a-Si:H solar cells grown using the ECR process*, in: Amorphous and Microcrystalline Silicon Technology - 1997, edited by S. Wagner, M. Hack, E.A. Schiff, R. Schropp, and I. Shimizu, Materials Research Society Symp. Proc. **467** (1997) 813-817.

Doyle, J., R. Robertson, G.H. Lin, M.Z. He, and A. Gallagher, *Production of high-quality amorphous silicon films by evaporative silane surface decomposition*, J. Appl. Phys. **64** (1988) 3215-3223.

Dusane, R.O., S.R. Dusane, V.G. Bhide, and S.T. Kshirsagar, *Hydrogenated microcrystalline silicon films produced at low temperature by the hot wire deposition method*, Appl. Phys. Lett. **63** (1993) 2201-2203.

Dutta, J., U. Kroll, P. Chabloz, A. Shah, A.A. Howling, J.-L. Dorier, and Ch. Hollenstein, *Dependence of intrinsic stress in hydrogenated amorphous silicon on excitation frequency in a plasma-enhanced chemical vapor deposition process*, J. Appl. Phys. **72** (1992) 3220-3222.

Faraji, M., S. Gokhale, S.M. Choudhari, M.G. Takwale, and S.V. Ghaisas, *High mobility hydrogenated and oxygenated microcrystalline silicon as a photosensitive material in photovoltaic applications*, Appl. Phys. Lett. **60** (1992) 3289-3291.

Feenstra, K.F., C.H.M. van der Werf, E.C. Molenbroek, and R.E.I. Schropp, *Deposition of device quality amorphous silicon by hot-wire CVD*, in: Amorphous and Microcrystalline Silicon Technology - 1997, edited by S. Wagner, M. Hack, E.A. Schiff, R. Schropp, and I. Shimizu, Materials Research Society Symp. Proc. **467** (1997) 645-650.

Feenstra, K.F.(Ph.D. Thesis, Utrecht University, The Netherlands, 1998).

Gallagher, A., *Neutral radical deposition from silane discharges*, J. Appl. Phys. **63** (1988) 2406-2413.

Ganguly, G., and A. Matsuda, *Importance of surface processes in defect formation in a-Si:H*, J. Non-Cryst. Solids **164-166** (1993) 31-36.

Ganguly, G., and A. Matsuda, *Role of hydrogen dilution in improvement of a-SiGe:H alloys*, J. Non-Cryst. Solids **198-200** (1996) 559-562.

Hamers, E.A.G., W.G.J.H.M. van Sark, J. Bezemer, H. Meiling, and W.F. van der Weg, *Structural properties of a-Si:H related to ion energy in VHF silane deposition plasmas*, J. Non-Cryst. Solids **226** (1998) 205-216.

Hattori, Y., D. Kruangam, K. Katoh, Y. Nitta, H. Okamoto and Y. Hamakawa, *High-conductivity wide band gap p-type a-SiC:H prepared by ECR-CVD and its application to high efficiency a-Si basis solar cells*, in: Proc. of the 19th IEEE PV Specialists Conf., 1987, 689-694.

Heintze, M., R. Zedlitz, and G.H. Bauer, *Mechanism of high rate a-Si:H deposition in a VHF plasma*, in: Amorphous Silicon Technology - 1993, edited by E.A. Schiff, M.J. Thompson, A. Madan, K. Tanaka, P.G. LeComber, Materials Research Society Symp. Proc. **297 297** (1993) 49-54.

Heintze, M., R. Zedlitz, H.N. Wanka, and M.B. Schubert, *Amorphous and microcrystalline silicon by hot wire chemical vapor deposition*, J. Appl. Phys. **79** (1996) 2699-2706.

Ichikawa, Y., K. Aizawa, H. Shimabukuro, Y. Nagao, and H. Sakai, *Deposition process and film properties of a-Si alloy films*, Technical Digest of the International PVSEC-3, Tokyo, Japan (1987) A-Ip-5.

Ishihara, S., D. He, M. Nakata, and I. Shimizu, *Preparation of high-quality microcrystalline silicon from fluorinated precursors by a layer-by-layer technique*, Jpn. J. Appl. Phys. **32** (1993) 1539-1545.

Ishihara, S., D. He, and I. Shimizu, *Structure of polycrystalline silicon thin film fabricated from fluorinated precursors by layer-by-layer technique*, Jpn. J. Appl. Phys. **33** (1994) 51-56.

Jansen, F., I. Chen, and M.A. Machonkin, *On the thermal dissociation of hydrogen*, J. Appl. Phys. **66** (1989) 5749-5755.

Jones, S.J., X. Deng, T. Liu, and M. Izu, *Preparation of a-Si:H and a-SiGe:H i-layers for nip solar cells at high deposition rates using a very high frequency technique*, in: Amorphous and Microcrystalline Silicon Technology - 1998, edited by R. Schropp, H. Branz, S. Wagner, M. Hack, and I. Shimizu, Materials Research Society Symp. Proc. **507** (1998) in print.

Keppner, H., P. Torres, J. Meier, R. Platz, D. Fischer, U. Kroll, S. Dubail, J.A. Anna Selvan, N. Pellaton Vaucher, Y. Ziegler, R. Tscharner, Ch. Hof, N. Beck, M. Goetz, P. Pernet, M. Goerlitzer, N. Wyrsch, J. Veuille, J. Cuperus, A. Shah, and J. Pohl, *The "Micromorph" cell: a new way to high-efficiency low-temperature crystalline silicon thin-film cell manufacturing?*, in: Advances in Microcrystalline and Nanocrystalline Semiconductors - 1996, edited by R.W. Collins, P.M. Faucher, I. Shimizu, J.C. Vial, T. Shimada, and A.P. Alivisatos, Materials Research Society Symp. Proc. **452** (1996) 865-876.

Kessels, W.M.M., M.C.M. van de Sanden, and D.C. Schram, *Hydrogen poor cationic silicon clusters in an expanding argon-hydrogen-silane plasma*, Appl. Phys. Lett. **72** (1998) 2397-2399.

Kinoshita, T., M. Shima, A. Terakawa, M. Isomura, H. Haku, K. Wakisaka, M. Tanaka, S. Kiyama, and S. Tsuda, *Effect of hydrogen dilution on a-Si/a-SiGe tandem solar cells*, Proceedings of the 14th European Photovoltaic Solar Energy Conference, June 30 - July 4, Barcelona, Spain, Eds. H.A. Ossenbrink, P. Helm, and H. Ehmann (H.S. Stephens and Associates, 1997) p566-569.

Kondo, M., Y. Toyoshima, A. Matsuda, and K. Ikuta, *Substrate dependence of initial growth of microcrystalline silicon in plasma-enhanced chemical vapor deposition*, J. Appl. Phys. **80** (1996) 6061-6063.

Kurihara, K., K. Sasaki, M. Kawarada, and N. Koshino, *High rate synthesis of diamond by dc plasma jet chemical vapor deposition*, Appl. Phys. Lett. **52** (1988) 437-438.

Luft, W. and Y. Simon Tsuo, *Hydrogenated amorphous silicon alloy deposition processes*, Applied Physics Series **1** (Marcel Dekker, Inc., New York, Basel, Hong Kong, 1993).

Madan, A., P. Rava, R.E.I. Schropp, and B. von Roedern, *A new modular multichamber plasma-enhanced chemical vapor deposition system*, Appl. Surf. Sci. **70/71** (1993) 716-721.

Mahan, A.H., and M. Vaněček, *A reduction in the Staebler-Wronski effect observed in low H content a-Si:H films deposited by the hot wire technique*, AIP Conf. Proc. **234** (1991) 195-202.

Mahan, A.H., J. Carapella, B.P. Nelson, R.S. Crandall, and I. Balberg, *Deposition of device quality, low H content amorphous silicon*, J. Appl. Phys. **69** (1991a) 6728-6730.

Mahan, A.H., B.P. Nelson, S. Salamon, and R.S. Crandall, *Deposition of device quality, low H content a-Si:H by the hot wire technique*, J. Non-Cryst. Solids **137 & 138** (1991b) 657-660.

Mahan, H., R.C. Reedy Jr., E. Iwaniczko, Q. Wang, B.P. Nelson, Y. Xu, A.C. Gallagher, H.M. Branz, R.S. Crandall, J. Yang, and S. Guha, *H out-diffusion and device performance in n-i-p solar cells utilizing high temperature hot wire a-Si:H i-layers*, in: Amorphous and Microcrystalline Silicon Technology - 1998, edited by R. Schropp, H. Branz, S. Wagner, M. Hack, and I. Shimizu, Materials Research Society Symp. Proc. **507** (1998) in print.

Matsuda, A., *Formation kinetics and control of microcrystallite in μc-Si:H from glow discharge plasma*, J. Non-Cryst. Solids **59-60** (1983) 767-774.

Matsumura, H., *Catalytic Chemical Vapor Deposition (CTL-CVD) method producing high quality hydrogenated amorphous silicon*, Jpn. J. Appl. Phys. **25** (1986) L949-L951.

Matsumura, H., *Catalytic chemical vapor deposition (CTL-CVD) method to obtain high quality amorphous silicon alloys*, in: Amorphous Silicon Technology, edited by A. Madan, M.J. Thompson, P.C. Taylor, P.G. LeComber, and Y. Hamakawa, Materials Research Society Symp. Proc. **118** (1988) 43-48.

Matsumura, H., *Study on catalytic chemical vapor deposition method to prepare hydrogenated amorphous silicon*, J. Appl. Phys. **65** (1989) 4396-4402.

Matsumura, H., A. Heya, R. Iizuka, A. Izumi, A.-Q. He, and N. Otsuka, *Low-temperature formation of device-quality polysilicon films by CAT-CVD method*, in: Advances in Microcrystalline and Nanocrystalline Semiconductors - 1996, edited by R.W. Collins, P.M. Faucher, I. Shimizu, J.C. Vial, T. Shimada, and A.P. Alivisatos, Materials Research Society Symp. Proc. **452** (1997) 982-988.

Meier, J., R. Flückiger, H. Keppner, and A. Shah, *Complete microcrystalline p-i-n solar cell - crystalline or amorphous cell behavior*, Appl. Phys. Lett. **65** (1994) 860-862.

Meier, J., S. Dubail, D. Fischer, J.A. Anna Selvan, N. Pellaton Vaucher, R. Platz, C. Hof, R. Flückiger, U. Kroll, N. Wyrsch, P. Torres, H. Keppner, A. Shah, and K.-D.

Ufert, *The "micromorph" solar cells: a new way to high efficiency thin film silicon solar cells*, Proceedings of the 13th EC Photovoltaic Solar Energy Conference, Edited by W. Freiesleben, W. Palz, H.A. Ossenbrink, and P. Helm (H.S. Stephens and Assoc., 1995) 1445.

Meier, J., H. Keppner, S. Dubail, U. Kroll, P. Torres, P. Pernet, Y. Ziegler, J.A. Anna Selvan, J. Cuperus, D. Fischer, and A. Shah, *Microcrystalline single-junction and micromorph tandem thin-film silicon solar cells*, in: Amorphous and Microcrystalline Silicon Technology - 1998, edited by R. Schropp, H. Branz, S. Wagner, M. Hack, and I. Shimizu, Materials Research Society Symp. Proc. **507** (1998) in print.

Meiling, H., M.J. van den Boogaard, R.E.I. Schropp, J. Bezemer, and W.F. van der Weg, *Hydrogen dilution of silane: Correlation between the structure and optical band gap in GD a-Si:H films*, in: Amorphous Silicon Technology - 1990, edited by P.C. Taylor, M.J. Thompson, P.G. LeComber, Y. Hamakawa, and A. Madan, Materials Research Society Symp. Proc. **192** (1990) 645-650.

Meiling, H., *Deposition of amorphous silicon thin films and solar cells*, Ph. D. Thesis (Utrecht University, Utrecht, The Netherlands, 1991).

Meiling, H., W.G.J.H.M. van Sark, J. Bezemer, and W.F. van der Weg, *Deposition-rate reduction through improper substrate-to-electrode attachment in very-high-frequency deposition of a-Si:H*, J. Appl. Phys. **80** (1996) 3546-3551.

Meiling, H., J. Bezemer, R.E.I. Schropp, and W.F. van der Weg, *High deposition rate a-Si:H through VHF-CVD of Ar diluted silane*, in: Amorphous and Microcrystalline Silicon Technology - 1997, edited by S. Wagner, M. Hack, E.A. Schiff, R. Schropp, and I. Shimizu, Materials Research Society Symp. Proc. **467** (1997a) 459-470.

Meiling, H., and R.E.I. Schropp, *Stable amorphous silicon thin film transistors*, Appl. Phys. Lett. **70** (1997b) 2681-2683.

Meiling, H., A.M. Brockhoff, J.K. Rath, and R.E.I. Schropp, *Transistors with a profiled active layer made by hot-wire CVD*, in: Amorphous and Microcrystalline Silicon Technology - 1998, edited by R. Schropp, H. Branz, S. Wagner, M. Hack, and I. Shimizu, Materials Research Society Symp. Proc. **507** (1998) in print.

Middya, A.R., J. Guillet, J. Perrin, A. Lloret and E. Bourree, *Hot-wire chemical vapour deposition of polycrystalline silicon films*, Proceedings of the 13th EC Photovoltaic Solar Energy Conference, edited by W. Freiesleben, W. Palz, H.A. Ossenbrink, and P. Helm (H.S. Stephens and Ass., Bedford, UK, 1995) 679-682.

Middya, A.R., S. Hazra, S. Ray, C. Longeaud, and J. P. Kleider, *a-Si:H and a-SiGe:H alloys fabricated close to powder regime of rf PECVD*, in: Amorphous and Microcrystalline Silicon Technology - 1997, edited by S. Wagner, M. Hack, E.A. Schiff, R. Schropp, and I. Shimizu, Materials Research Society Symp. Proc. **467** (1997) 615-620.

Molenbroek, E.C., *Deposition of hydrogenated amorphous silicon with the hot-wire technique* (Ph.D. Thesis, University of Colorado, 1995a).

Molenbroek, E.C., A.H. Mahan, E.J. Johnson, and A.C. Gallagher, *Film quality in relation to deposition conditions of a-Si:H films deposited by the "hot wire" method using highly diluted silane*, J. Appl. Phys. **79** (1996) 7278-7292.

Molenbroek, E.C., A.H. Mahan, and A. Gallagher, *Mechanisms influencing "hot-wire" deposition of hydrogenated amorphous silicon*, J. Appl. Phys. **82** (1997) 1909-1917.

Nelson, B.P., E. Iwaniczko, R.E.I. Schropp, H. Mahan, E. Molenbroek, S. Salamon, and R.S. Crandall, *Amorphous silicon solar cells incorporating hot-wire deposited intrinsic material*, Proceedings of the 12th International E.C. Photovoltaic Solar Energy Conference 1994, Edited by R. Hill, W. Palz, and P. Helm, (H.S. Stephens and Associates, 1994) 679-682.

Oda, S., J. Noda, and M. Matsumura, *Preparation of a-Si:H films by VHF plasma CVD*, in: Amorphous Silicon Technology, edited by A. Madan, M.J. Thompson, P.C. Taylor, P.G. LeComber, and Y. Hamakawa, Materials Research Society Symp. Proc. **118** (1988) 117-122.

Okamoto, S., Y. Hishikawa, and S. Tsuda, *New interpretation of the effect of hydrogen dilution of silane on glow-discharged hydrogenated amorphous silicon for stable solar cells*, Jpn. J. Appl. Phys. **35** (1996) 26-33.

Papadopoulos, P., A. Scholz, S. Bauer, B. Schröder, and H. Öchsner, *Deposition of device quality a-Si:H films with the hot-wire technique*, J. Non-Cryst. Solids **164-166** (1993) 87-90.

Perrin, J., Y. Takeda, N. Hirano, Y. Takeuchi, and A. Matsuda, *Sticking and recombination of the SiH_3 radical on hydrogenated amorphous silicon: the catalytic effect of diborane*, Surf. Sci. **210** (1989) 114-128.

Perrin, J., and J. Schmitt, *Diagnostics and modelling of rf plasma deposition of amorphous silicon and powder formation in large area symmetric parallel-plate reactors*, 11th E.C. Photovoltaic Solar Energy Conference 1992, Eds. L. Guimarães, W. Palz, C. de Reyff, H. Kiess, and P. Helm (Harwood Academic Publishers, 1992) 80-83.

Perrin, J., in: *Plasma deposition of amorphous-based materials*, eds. G. Bruno, P. Capezzuto, and A. Madan (Academic Press Inc., San Diego, CA, 1995) 216.

Rech, B., C. Beneking, S. Wieder, Th. Eickhoff, and H. Wagner, *Development of a-Si:H/a-Si:H stacked solar cells with high efficiency and high light stability*, Proceedings of the 13th EC Photovoltaic Solar Energy Conference, Edited by W. Freiesleben, W. Palz, H.A. Ossenbrink, and P. Helm (H.S. Stephens and Assoc., 1995) 613.

Rath, J.K., M. Galetto, C.H.M. van der Werf, K. Feenstra, H. Meiling, M.W.M. van Cleef, and R.E.I. Schropp, *Hot Wire CVD: A one-step process to obtain thin film polycrystalline silicon at a low temperature on cheap substrates*, Technical Digest of the 9th International Photovoltaic Science and Engineering Conference, Nov. 11-15, 1996, Miyazaki, Japan, p227;
J.K. Rath, H. Meiling, and R.E.I. Schropp, *Low-temperature deposition of polycrystalline silicon thin films by hot-wire CVD*, Solar Energy Materials and Solar Cells **48** (1997) 269-277.

Rath, J.K., A.J.M.M. van Zutphen, H. Meiling, and R.E.I. Schropp, *Application of hot wire deposited intrinsic poly-silicon films in n-i-p cells and TFTs*, in: Amorphous and Microcrystalline Silicon Technology - 1997, edited by S. Wagner, M. Hack, E.A. Schiff, R. Schropp, and I. Shimizu, Materials Research Society Symp. Proc. **467** (1997) 445-450.

Rath, J.K., K.F. Feenstra, C.H.M. van der Werf, Z. Hartman, and R.E.I. Schropp, *Profiled hot-wire CVD poly-Si:H films for an n-i-p cell on a metal substrate*, presented at the 2nd World Conference and Exhibition on Photovoltaic Energy Conversion, Vienna, July 1998.

Schmitt, J.P.M., J. Meot, P. Roubeau, and P. Parrens, *New reactor design for low contamination amorphous silicon deposition*, Proceedings of the 8th European Community Photovoltaic Solar Energy Conference, Florence 1988, Editors: I. Solomon, B. Equer, and P. Helm (Kluwer Academic Publishers, Dordrecht/Boston/London, The Netherlands, 1988) 964.

Schropp, R.E.I., B. von Roedern, P. Klose, R.E. Hollingsworth, J. Xi, J. del Cueto, H. Chatham, and P.K. Bhat, *Recent progress in multichamber deposition of high quality amorphous silicon solar cells on planar and compound curved substrates at GSI*, Solar Cells **27** (1989) 59-68.

Schropp, R.E.I., K.F. Feenstra, E.C. Molenbroek, H. Meiling and J.K. Rath, *Device-quality polycrystalline and amorphous silicon films by Hot Wire Chemical Vapor Deposition*, Phil. Mag. B **76** (1997) 309-321.

Severens, R.J., M.C.M. van de Sanden, H.J.M. Verhoeven, J. Bastiaanssen, and D.C. Schram, *On the effect of substrate temperature on a-Si:H deposition using an expanding thermal plasma*, in: Amorphous Silicon Technology - 1996, edited by M. Hack, E.A. Schiff, S. Wagner, R. Schropp, and A. Matsuda, Materials Research Society Symp. Proc. **420** (1997) 341-346.

Shibata, N., K. Fukuda, H. Ohtoshi, J. Hanna, S. Oda, and I. Shimizu, *Growth of amorphous and crystalline silicon by HR-CVD*, in: Amorphous Silicon Semiconductors - Pure and Hydrogenated, edited by A. Madan, M. Thompson, D. Adler, and Y. Hamakawa, Materials Research Society Symp. Proc. **95** (1987) 225-235.

Shing, Y.H., *Electron cyclotron resonance deposition and plasma diagnostics of a-Si:H and a-C:H films*, Solar Cells **27** (1989) 331-340.

Shirafuji, J., S. Nagata, and M. Kuwagaki, *Effect of hydrogen dilution of silane on optoelectronic properties in glow-discharged hydrogenated silicon films*, J. Appl. Phys. **58** (1985) 3661-3663.

Shirai, H., D. Das, J. Hanna, and I. Shimizu, *A novel preparation technique for preparing hydrogenated amorphous silicon with a more rigid and stable Si network*, Appl. Phys. Lett. **59** (1991) 1096-1098

Shirai, H., B. Drévillon, and I. Shimizu, *Role of hydrogen plasma during growth of hydrogenated microcrystalline silicon - In situ UV-visible and infrared ellipsometry study*, Jpn. J. Appl. Phys. **33** (1994) 5590-5598.

Sterling, H.F. and R.C.G. Swann, *Chemical Vapour deposition promoted by r.f. discharge*, Solid-State Electron. **8** (1965) 653-654.

Tsai, C.C., R. Thompson, C. Doland, F.A. Ponce, G.B. Anderson and B. Wacker, *Transition from amorphous to crystalline silicon: effect of hydrogen on film growth*, in: Amorphous Silicon Technology, edited by A. Madan, M.J. Thompson, P.C. Taylor, P.G. LeComber, and Y. Hamakawa, Materials Research Society Symp. Proc. **118** (1988) 49-54.

Vepřek, S., F.A. Sarrott, S. Rambert, and E. Taglauer, *Surface hydrogen content and passivation of silicon deposited by plasma induced chemical vapor deposition from silane and the implications for the reaction mechanism*, J. Vac. Sci. Technol. A **7** (1989) 2614-2624.

Watanabe, T., M. Tanaka, K. Azuma, M. Nakatani, T. Sonobe, and T. Shimada, *Chemical Vapor Deposition of a-SiGe:H films utilizing a microwave excited plasma*, Jpn. J. Appl. Phys. **26** (1987) L288-L290.

Wiesmann, H., A.K. Ghosh, T. McMahon, and M. Strongin, J. Appl. Phys. **50** (1979) 3752; H.J. Wiesmann, *Method of producing hydrogenated amorphous silicon film*, U.S. Patent No. 4,237,150; Dec. 2, 1980.

Xi, J.P., R.E. Hollingsworth, P.K. Bhat, and A. Madan, *Deposition rate effects on amorphous silicon solar cell stability*, AIP Conf. Proc. **157** (1987) 158-164.

Xu, X., J. Yang, and S. Guha, *Hydrogen dilution effects on a-Si:H and a-SiGe:H materials properties and solar cell performance*, J. Non-Cryst. Solids **198-200** (1996) 60-64.

Xu, X., J. Jang, and S. Guha, *On the lack of correlation between film properties and solar cell performance of amorphous silicon-germanium alloys*, Appl. Phys. Lett. **62** (1993) 1399-1401.

Yamamoto, K., *Thin film poly-Si solar cell on glass substrate fabricated at low temperature*, in: Amorphous and Microcrystalline Silicon Technology - 1998, edited by R. Schropp, H. Branz, S. Wagner, M. Hack, and I. Shimizu, Materials Research Society Symp. Proc. **507** (1998) in print.

Yang, L., L. Chen, and A. Catalano, *The effect of hydrogen dilution on the deposition of SiGe alloys and the device stability*, in: Amorphous Silicon Technology - 1991, edited by A. Madan, Y. Hamakawa, M.J. Thompson, P.C. Taylor, and P.G. LeComber, Materials Research Society Symp. Proc. **219** (1991) 259-264.

Zafar, S., and E.A. Schiff, *Hydrogen-mediated model for defect metastability in hydrogenated amorphous silicon*, Phys. Rev. B **40** (1989) 5235-5238.

Zedlitz, R., F. Kessler, and M. Heintze, *Deposition of a-Si:H with the hot-wire technique*, J. Non-Cryst. Solids, **164-166** (1993) 83-86.

3 OPTICAL, ELECTRONIC AND STRUCTURAL PROPERTIES

This research is not purely academic: disordered phases of condensed matter – steel and glass, earth and water, if not fire and air – are far more abundant, and of no less technological value, than the idealized single crystals that used to be the sole object of study of "solid state physics".

—J.M. Ziman, 1979

3.1 UNDOPED MATERIALS

3.1.1 Hydrogenated amorphous silicon

In the literature intrinsic plasma-deposited amorphous silicon prepared under certain conditions is often denoted as *device-quality* material. Obviously, the criteria for device quality depend on the specific device and also on which part of the device the material is used for. Limiting ourselves to solar cells, we will describe in this chapter what exactly is meant by *device quality*, and expand this definition to thin-film micro- and polycrystalline silicon as well.

The minimum set of characteristic parameters that should be determined to qualify intrinsic hydrogenated amorphous silicon for use in p-i-n type solar cells consists of the dark conductivity $\sigma_{\rm d}$, the photoconductivity $\sigma_{\rm ph}$, the optical band gap E_g, and the activation energy $E_{\rm A}$ of the dark conductivity. The absorption coefficients at the wavelengths of 400 nm and 600 nm are also useful to specify the quality of the material. Additional parameters are the midgap electronic density of states and the hydrogen bonding structure parameters as obtained from infrared (IR) absorption spectra.

The dark conductivity $\sigma_{\rm d}$. This can be determined using a single layer on highly resistive glass, e.g., Corning 7059 or equivalent (Corning 1737F). Most simply, two (> 1 cm long) coplanar strips of a metal with a low work function (e.g., silver) are evaporated or pasted less than 1 mm apart on top of the layer, each providing an ohmic contact. The layer under study should be thick enough (i.e. $\approx 1\ \mu$m) to prevent depletion extending in a region larger than the contact region. Using a picoampere meter and a metal box providing a grounded shield and preventing all traces of stray light $\sigma_{\rm d}$ can be determined using voltages of typically 100 V, from

$$\sigma_{\rm d} = \frac{Iw}{Vld} \tag{3.1}$$

where I is the measured current, V the applied voltage, d the thickness of the silicon film, w the distance between the contacts, and l the length of the contacts. For device quality material, $\sigma_{\rm d} < 1 \times 10^{-10}\ \Omega^{-1}{\rm cm}^{-1}$.

The photoconductivity $\sigma_{\rm ph}$. This is determined using the same electrode geometry while the film is illuminated with light that has the AM1.5 spectrum, at an intensity of 100 mW/cm^2, preferably from a solar simulator. As this is wide-spectrum illumination the film should not be much thicker than 1 μm, in order to photoexcite the entire thickness of the film. The requirement here is that $\sigma_{\rm ph} > 1 \times 10^{-5}\ \Omega^{-1}{\rm cm}^{-1}$.

The photoconductivity $\sigma_{\rm ph}$ can be expressed as (Zanzucchi et al., 1977):

$$\sigma_{\rm ph} = e\mu\Delta n = \frac{e\eta_g\mu\tau(1-R_f)F_0}{d}[1-\exp(-\alpha d)] \tag{3.2}$$

where Δn is the density of photogenerated majority carriers (electrons), μ their mobility, τ their lifetime, η_g the quantum efficiency for carrier generation, R_f the reflectance of the film, d its thickness, F_0 the illumination intensity measured in photons per cm^2 per second, and α the absorption coefficient of the material at the probe wavelength.

Therefore an additional useful figure of merit that includes photoabsorption, reflection, thickness of the film, carrier transport and recombination is the quantum efficiency-mobility-lifetime product, $\eta_g\mu\tau$. This quantity is best measured using a relatively long wavelength at which the absorption coefficient is small in order to ascertain near-uniform carrier generation throughout the thickness of the film. For instance, λ = 600 nm can be chosen as the probe wavelength.

Using Eqn. 3.2 combined with geometry factors as in Eqn. 3.1, one obtains for this specific wavelength

$$(\eta_g \mu \tau)_{600} = \frac{Iw}{e(1 - R_f)F_0 V l[1 - \exp(-\alpha_{600} d)]} \tag{3.3}$$

In amorphous silicon, $\eta_g = 1$ for *absorbed* photons. A *device-quality* value for the mobility-lifetime product at 600 nm is $\mu\tau \geq 1 \times 10^{-7}$ cm^2/V.

Optical band gap E_g. The optical band gap can be determined from spectral optical transmission measurements, or more accurately from combined reflection and transmission measurements.. Due to the lack of translational symmetry in amorphous semiconductors the law of momentum conservation is relaxed so that amorphous silicon behaves like a direct gap semiconductor. Therefore the optical absorption coefficient α is determined by the availability of electronic states only (Tauc, 1972). The optical band gap is determined by extrapolating $[\alpha(E)n(E)E]^{1/1+p+q}$ versus the photon energy E to $\alpha(E) = 0$, for $\alpha \geq 10^3$ cm^{-1}:

$$(\alpha(E)n(E)E)^{1/1+p+q} = B_g \cdot (E - E_g) \tag{3.4}$$

where $\alpha(E)$ is the absorption coefficient, $n(E)$ the refractive index, p and q are constants related to the shape of the band edges and B_g is a prefactor. If the density of states has a square-root energy dependence near the band edges, as is commonly the case in crystalline semiconductors (p=q=1/2), Eqn. 3.4 describes the so-called Tauc plot and the corresponding gap the Tauc gap. If the distribution near the band edges is assumed to be linear (p=q=1), as proposed by Klazes et al., 1982, E_g is called the cubic gap. The value of the cubic gap is 0.1-0.2 eV lower than that of the Tauc gap. We stress that the fitting procedure depends on the energy range that is chosen for the fit. The cubic plot of Eqn. 3.4 is linear in a larger energy interval than the Tauc plot.

A judiciously chosen fit to data taken from intrinsic a-Si:H should yield $E_g <$ 1.80 eV for the Tauc gap and $E_g <$ 1.60 eV for the cubic gap. The absorption coefficient at 600 nm is preferably $\alpha_{600} > 3.5 \times 10^4$ cm^{-1} and at 400 nm it is $\alpha_{400} > 5 \times 10^5$ cm^{-1}.

The conduction and valence band tail states give rise to subbandgap absorption, which can be fitted to $\alpha = \alpha_0 \exp(E/E_{V0}^{tail})$ where E_{V0}^{tail} is the Urbach energy. The Urbach energy reflects the slope of the exponential region of the valence band tail. A typical value for device quality material is $E_{V0}^{tail} \approx 50$ meV.

The activation energy E_A. The photoresponse σ_{ph}/σ_d reflects the opto-electronic quality to a large extent, but is itself relatively independent of the position of the Fermi level. A low dark conductivity $\sigma_d < 1 \times 10^{10}\ \Omega^{-1}cm^{-1}$ does not guarantee the absence of electronically active impurities as it could just as well be due to a relatively large band gap.

The activation energy E_A of the dark conductivity is a good measure of the energy difference between the Fermi level and the conduction band edge for electron transport (the valence band edge for hole transport). It is obtained from the temperature dependent conductivity $\sigma(T)$ by fitting to the relation:

$$\sigma(T) = \sigma_0 \exp(-E_A/kT) \tag{3.5}$$

in which σ_0 is a conductivity prefactor, T the absolute temperature and k Boltzmann's constant. This linear relationship between $\log(\sigma(T))$ and $1/T$ is an approximation since the mobility, and thus the prefactor, is weakly temperature dependent. Apart from this, σ_0 as determined for a variety of materials shows a correlation with E_A according to the Meyer-Neldel rule (Meyer and Neldel, 1937). The origin of this correlation is not well established, but is likely to be due to the difference in temperature dependence of the band-gap energy, Fermi-level position, and carrier transport energy level for different samples (Fritzsche and Tanielian, 1981). Therefore, E_A does not exactly quantify equivalent energy separations for different materials. Stutzmann, 1987 estimated that E_A can be determined in the temperature range $50° < T < 160°$ C with an accuracy of $\approx$ 100 meV.

Combined with the optical band gap, the value of E_A is a very accurate indicator for the presence or absence of impurities. Small concentrations (1×10^{16} cm^{-3}) of B or P can shift the Fermi level over several tenths of eV. Also concentrations of 1×10^{17} cm^{-3} of O or N have this effect, thus severely limiting the minority carrier drift/diffusion length in devices. Only in the event that oxygen impurities (n-type dopant) are compensated by boron impurities (p-type dopant), the activation energy values can be misinterpreted as representing high quality material.

For a truly intrinsic material the Fermi level should be at the middle of the band gap. Many laboratories find that the Fermi level is slightly above the midgap position, which is often interpreted as n-type behaviour. However, n-type behaviour in *intrinsic* material originates from the higher mobility of electrons compared to holes whereas unintentionally incorporated impurities, such as oxygen, effectively dope the material causing an upward shift of the Fermi level. A good value for E_A for undoped a-Si:H is $\approx$ 800 meV, which is roughly half the *cubic* band gap.

The midgap density of states. Many methods exist for the determination of the density of defect states distributions within the band gap. Some rely on electrical measurements of a semiconductor/metal junction, such as DLTS (Deep Level Transient Spectroscopy, Lang et al., 1982) and ICTS (Isothermal Capacitance Transient Spectroscopy, Jackson et al., 1985), or of a semiconductor/insulator junction, such as the field effect measurement. Other methods use a single layer with coplanar contacts, such as the CPM (Constant Photocurrent Measurement; Moddel et al., 1980) and the TSC (Thermally Stimulated Conductivity; Zhu and Fritzsche, 1986) methods. A few methods, such as SCLC (Space Charge Limited Current; Den Boer, 1981, Mackenzie et al., 1982) and DBP (Dual Beam Photoconductivity) can be applied to sandwich devices. These methods usually lead to smaller values for the density of states, presumably because they are less influenced by surface effects. PDS (Photothermal Deflection Spectroscopy; Jackson et al., 1981) is a contactless method with a high sensitivity, but quite surface sensitive. ESR (Electron Spin Resonance; Kumeda and Shimizu, 1980) (or EPR, Electron Paramagnetic Resonance) is perhaps the only true bulk defect measurement, but it underestimates the defect density as it only detects the paramagnetic spin centers. Charged dangling bonds are not detected as they do not possess an unpaired spin signal.

Here it is difficult to give a unique value for the density of defect states N(E), below which the material can be considered *device quality*. For surface sensitive methods, a rule of thumb is that the maximum in the dangling bond distribution is smaller than 2×10^{16} $cm^{-3}eV^{-1}$. For more truly 'bulk'-type methods, this should be $< 1 \times 10^{16}$ $cm^{-3}eV^{-1}$. Using ESR, the spin density should be $N_s < 8 \times 10^{15}$ cm^{-3}.

The hydrogen content. The hydrogen content can be determined from SIMS (Secondary Ion Mass Spectroscopy), ERD (Elastic Recoil Detection), NRA (Nuclear Reaction Analysis), or from hydrogen evolution measurements. An experimentally more convenient method, and therefore the most commonly used, is FTIR (Fourier-transformed Infrared Transmittance) spectroscopy.

In this indirect method for the determination of the hydrogen content it is important to accurately correct for optical interference and background signals. To eliminate the interference effect, the spectra can be corrected using a procedure proposed by Brodsky et al., 1977. This correction assumes that the refractive indices of the film (a-Si:H) and the substrate (c-Si) are equal and therefore the interference occurs in the film and the substrate together. In reality, the refractive index of a-Si:H depends on the hydrogen content and differs from the refractive index of c-Si. Therefore, interference also occurs in the a-Si:H film. It has been shown by Maley, 1992 and Tzolov et al., 1993 that the

Brodsky correction leads to considerable overestimate of the absorption data for films thinner than $\approx 1\ \mu m$.

If the investigated films are thinner than 1 μm, the Brodsky correction has to be corrected itself by additional correction factors listed by Maley, 1992 to compensate for the overestimate of the IR absorption. If the films are thicker than 1 μm, the Brodsky correction itself is adequate. To account for the absorption of the c-Si substrates, the measured spectrum needs to be corrected with a background spectrum of the substrate with no film.

The hydrogen content can most unambiguously be determined from the absorption peak at 640 cm^{-1}, which includes the rocking mode of bonded hydrogen in every possible bonding configuration (i.e. mono-, di-, and trihydride bonding as well as in polymeric chains).

The absorption at 640 cm^{-1} can be fitted with a single gaussian to obtain the absorption strength of the rocking mode $\alpha_{640}(\omega)$. The integrated absorption coefficient I_{640} is defined by:

$$I_{640} = \int_{-\infty}^{+\infty} \alpha_{640}(\omega) d\omega/\omega \tag{3.6}$$

The hydrogen concentration [H] is proportional to the integrated absorption coefficient

$$[H] = A_{640} I_{640} \tag{3.7}$$

The value of $A_{640} = 2.1 \times 10^{19}$ cm^{-2} is currently most widely used for the determination of the hydrogen content from the rocking mode. The determination of the hydrogen content from the stretching modes at 2000 - 2090 cm^{-1}, has been a matter of discussion due to the possibility of making different assignments to these modes and the concomitant uncertainty in the proportionality factors (Langford et al., 1992, Daey Ouwens and Schropp, 1996). However, it has been confirmed by Beyer and Abo Ghazala, 1998 that the value of the proportionality constant is independent of the position of the stretching mode in the region 2000 - 2090 cm^{-1}, and amounts to $A_{2000-2090} = 1.1 \times 10^{20}$ cm^{-2}. It further appeared that both A_{640} and $A_{2000-2090}$ can be weakly dependent on the total hydrogen concentration. The values quoted here are appropriate for samples with a total hydrogen concentration of about 10 at.-%.

For PECVD materials it is known that a hydrogen content below 8 at.-% generally leads to a too high density of defect states and a reduced photoresponse, and that a hydrogen content higher than 12 % leads to the formation of dihydride bonds and inclusions of clustered hydrogen, which is often accompanied by an increased density of states. For materials deposited otherwise, e.g., by ECR-CVD or HWCVD, it is possible to obtain device-quality properties at a hydrogen concentration < 8 at.-%, and even down to 0.07 at.-% in the case of HWCVD (Mahan et al., 1991).

It is often convenient to define a 'microstructure parameter', denoted as R^*, as a figure of merit indicating the fraction of hydrogen atoms that are *not* in isolated SiH bonds surrounded by a dense network. Hydrogen bonded otherwise (i.e. monohydride bonds at a 'free' surface, e.g., a void, dihydride bonds, clustered monohydride or dihydride bonds) does not have a stretching mode vibration at 2000 cm^{-1}, but one that is shifted over 60 - 100 cm^{-1} to higher wavenumbers. The microstructure parameter is thus defined as

$$R^{\star} = \frac{I_{2000}}{I_{2000} + I_{2060-2100}} \tag{3.8}$$

Ideally, the value of $R^{\star}$ is zero, but for practical purposes $R^{\star} < 0.1$ is acceptable.

Summarizing, the criteria for device-quality intrinsic amorphous silicon are listed in Table 3.1.

Table 3.1. Criteria for 'device quality' amorphous silicon films.

	property	*requirement*
Minimum set	Dark conductivity	$< 1 \times 10^{-10}\ \Omega^{-1}\text{cm}^{-1}$
	AM1.5 100 mW/cm^2 photoconductivity	$> 1 \times 10^{-5}\ \Omega^{-1}\text{cm}^{-1}$
	Band gap, Tauc	< 1.8 eV
	Band gap, cubic	< 1.6 eV
	Absorption coefficient at 600 nm	$\geq 3.5 \times 10^4\ \text{cm}^{-1}$
	Absorption coefficient at 400 nm	$\geq 5 \times 10^5\ \text{cm}^{-1}$
	Activation energy	≈ 0.8 eV
Additional Characteristics	Density of dangling bond states	
	(Opto-) electrical methods	$\leq 1 \times 10^{16}\ \text{cm}^{-3}$
	Density of defects (ESR method)	$\leq 8 \times 10^{15}\ \text{cm}^{-3}$
	Hydrogen content (in PECVD material)	9-11 at.-%
	Microstructure parameter R^*	< 0.1
	Mobility-lifetime product at 600 nm	$\geq 1 \times 10^{-7}\ \text{cm}^2/\text{V}$

3.1.2 Micro- and polycrystalline silicon

This section deals with the properties of *undoped* micro- or polycrystalline silicon required for utilization in a p^+-i-n^+ or an n^+-i-p^+ device. The first application of microcrystalline silicon however was in the n- or p-type doped layers, which will be discussed in section 3.2.2. For successful application of microcrystalline silicon as the active (intrinsic) layer in a sandwich device, the primary requirement is that it contains a virtually no amorphous network tissue and therefore in this book it is preferably referred to as polycrystalline silicon (see table 1.1).

Suitable thin film intrinsic polycrystalline silicon can for instance be prepared by VHFCVD, HWCVD, or by PECVD at 13.56 MHZ if the layer-by-layer technique is used or similar methods known as *chemical annealing* (see section 2.2.1). The set of parameters for this type of material to qualify for use as the i-layer in solar cells is different from that for hydrogenated amorphous silicon, mainly because polycrystalline silicon is not an isotropic material. Thus it is important to determine structural properties related to the predominant direction for carrier transport, i.e. the size and orientation of the grains, the nature of the grain boundaries including hydrogen passivation of grain boundary defects.

The minimum set of characteristic parameters to be determined to qualify intrinsic poly-Si:H as *device-quality* are the crystalline volume fraction V_f, the orientation of the grains, the band gap E_g, the photoconductivity σ_{ph}, the dark conductivity σ_{d} and its activation energy E_{A}.

Additional parameters supporting the quality assessment are the hydrogen bonding configurations as detected by infrared absorption, the Hall mobility activation energy, the minority carrier diffusion length, and the majority carrier $\mu\tau$ -product. The defect density N_s is an important parameter, but it is the last one to be discussed since the value determined experimentally does not by itself decide whether a material is device quality or not. Here, the spatial distribution defines whether electronic defects adversely affect solar cell performance or not (Rath et al., 1997b).

It should be noted that for these genuine thin film devices the grain size is of minor importance, in contrast to general belief. Here one should distinguish between the high temperature approach (T > 700°C) for depositing polycrystalline silicon (such as LPE (Weber et al., 1995) or High Temperature CVD (Brendel et al., 1995)) requiring grains with a size > 50 μm, and the low temperature approach (T < 550°C) as discussed in this section. In the low temperature approach, the processing conditions for devices are compatible with materials with a low temperature resistance, such as glass. At first sight, it is surprising that polycrystalline films with a grain size that is sometimes even smaller than 20 nm can be used as the active layer in a p^+-i-n^+ solar cell. However, due to the low processing temperature, as-deposited films have a sufficient amount of natively built-in hydrogen so that grain boundaries are passivated very well. The high temperature approach, which is beyond the scope of this book, often requires a post-hydrogenation step to passivate the grain boundaries.

The crystalline volume fraction V_f. This parameter can most quickly be found by laser Raman spectroscopy. The Raman spectrum for single-crystal silicon shows a sharp peak at 520 cm^{-1}, associated with the transverse optical (TO) mode, whereas the amorphous silicon/grain boundary signal appears at 480 - 500 cm^{-1}. The crystalline volume fraction can be found from the deconvoluted intensities I_c and I_a of the Raman spectra, at 520 cm^{-1} and 480 - 500 cm^{-1}, respectively, by applying

$$V_f = \frac{I_c}{I_c + mI_a}, \tag{3.9}$$

where m is a correction for the crystallite size x, as proposed by Bustarret et al., 1988, which reads

$$m(x) = 0.1 + \mathrm{e}^{-x/250} \tag{3.10}$$

when x is the dimension of the grains in Å. The grain size can be obtained from X-ray diffraction (XRD) or under certain conditions from the width of the TO peak at 520 cm^{-1}. Good quality poly-Si:H has $V_f > 90$ %, although cells have been made with a lower V_f material. For materials with a grain size < 100 nm, V_f will never be completely 100 % as the Raman signal from grain boundaries will always yield an additional small signal shifted from the 520 cm^{-1} towards lower wavenumbers. As can be found in a publication by Vepřek et al., 1987, this signal appears at 505 cm^{-1}. It has been attributed to bond extension in essentially monatomic grain boundaries. The absence of a peak centered at 480 cm^{-1} may be regarded as evidence that a film contains no amorphous tissue and is truly polycrystalline. Fig. 3.1 shows typical deconvolutions of spectra from micro- and polycrystalline films.

The orientation and size of the grains. This property can be determined from X-ray diffraction (XRD) spectra. The diffraction pattern may reveal predominantly (220) orientation, but in addition it may show (111), (311), (331) and to a lesser extent (400), (422), (511). Ideally, the film has strongly anisotropic columnar orientation with only the (220) signal present. Rath et al., 1997 succeeded in obtaining this using the HWCVD deposition technique. The crystallite size x can be estimated from XRD by using the Scherrer formula (Klug and Alexander, 1974):

$$x = \frac{k\lambda}{(\Delta\theta)\cos\theta} \tag{3.11}$$

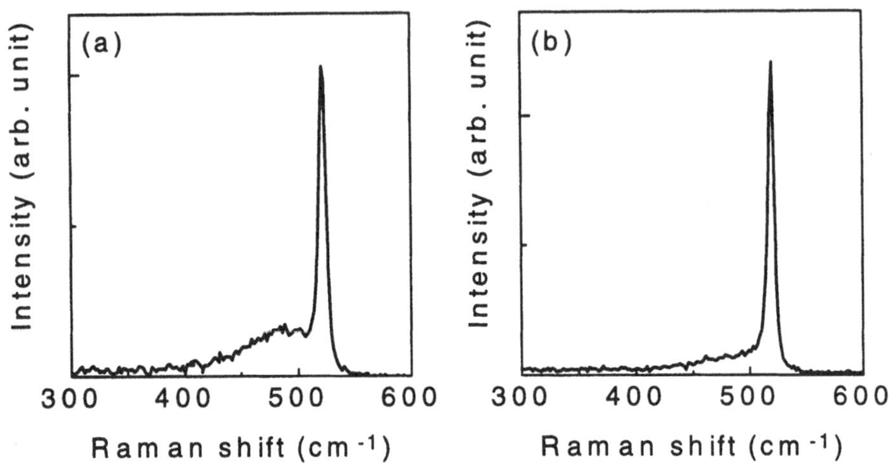

Figure 3.1. Typical Raman scattering spectra for (a) a mixed phase, or microcrystalline film, (b) a purely polycrystalline film.

Note that this formula applies to crystallites with a spherical shape only. As can be clearly seen in the Transmission Electron Microscopy (TEM) picture in Fig. 3.2, the films with a preferential (220) orientation have grains with a long columnar needle type shape, so that the Scherrer analysis will considerably underestimate the grain size. Therefore, only cross-sectional microscopy can complete the assessment of the structural morphology.

The crystallite size can also be obtained from the full width of half maximum (FWHM) of the TO peak, as outlined by Iqbal et al., 1981, however this analysis is only applicable for grain sizes up to 20 nm (Fauchet and Campbell, 1988).

The band gap E_g, the photo- and dark conductivity $\sigma_{\rm ph}$ and $\sigma_{\rm d}$, and the activation energy $E_{\rm A}$. The optical band gap for single crystal silicon is 1.1 eV. It is an indirect semiconductor gap, so that the absorption coefficient stays low up to photon energies high enough ($\leq$ 3.1 eV) to allow direct optical transitions. Nevertheless, the poly-Si:H solar cell benefits from absorption of photons with energies in the range of 1.1 eV - 1.75 eV, where an a-Si:H cell shows no response, and therefore poly-Si:H cells have an extended response in the near infrared part of the spectrum. The absorption spectra of various materials are compared in Fig. 3.3.

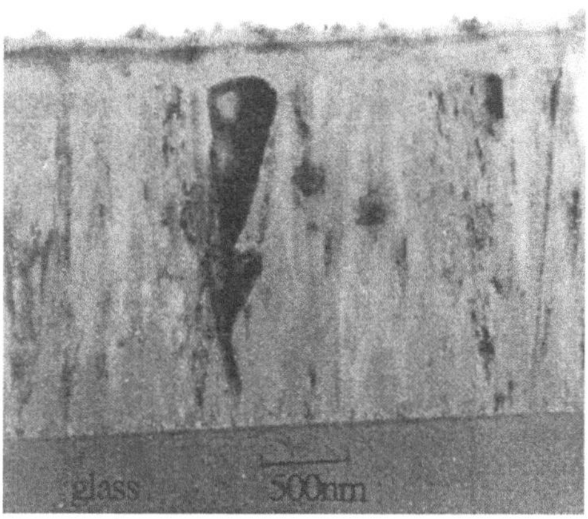

Figure 3.2 Cross-sectional TEM of a polycrystalline thin film that exhibits only (220) orientation in XRD spectra. To provide immediate nucleation on glass, the first 20 nm of the film was made under high hydrogen dilution conditions.

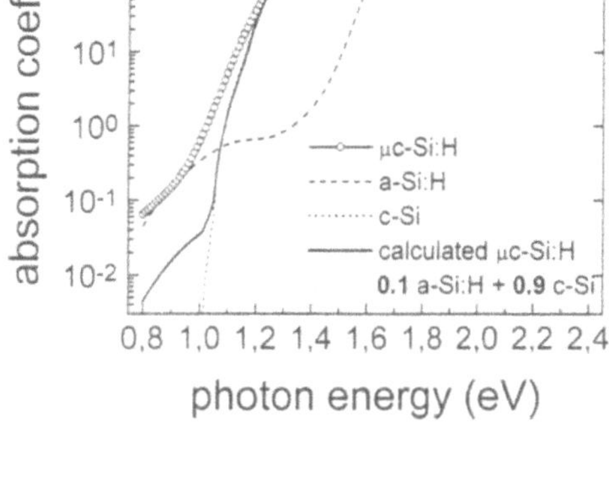

Figure 3.3 Comparison of the optical absorption coefficients of single crystal silicon, μc-S:H, and a-Si:H. Also shown is a curve calculated from the effective media approximation (Ward, 1994), using 10 % a-Si:H and 90 % crystalline silicon. (reprinted from Vaněček et al, 1997, with permission from Elsevier Science).

For an indirect-gap semiconductor the standard relation between the absorption coefficient $\alpha(E)$ and the photon energy E holds:

$$\alpha(E) = C \cdot (E - E_g \pm E_{ph})^2 \tag{3.12}$$

where E_{ph} is the phonon energy involved in an indirect transition and C is a prefactor which depends on the phonon absorption or emission rate during an indirect electron transition. Therefore, C is not a constant since it depends on the matrix element for electron-phonon interaction and the phonon density of states. Nevertheless, the band gap E_g can be estimated from a plot of $\sqrt{\alpha}$ versus the photon energy E, where the absorption coefficients are determined from PDS, CPM or DBP measurements. For polycrystalline films with a crystalline volume fraction V_f in excess of 90 %, values for E_g have been found between 1.0 eV (Beck et al., 1996) and 1.1 (Rath et al., 1997). If values larger than 1.1 eV are found, e.g., 1.3 - 1.4 eV (or in excess of 1.9 eV if the Tauc plot is used), then a significant fraction of amorphous content is present. This is typical for the early microcrystalline films made with conventional 13.56 MHz PECVD in a high power, high dilution regime (Matsuda, 1983).

The band-gap values lower than that of single crystal silicon have been attributed to tensile strain and optical scattering. All micro- or polycrystalline silicon films with grain sizes in the 1 - 1000 nm range show enhanced absorption as compared to crystalline silicon, irrespective of surface texture (recent data tables for c-Si were published by Green and Keevers, 1995). There has been some speculation as to the origin of this effect. If $V_f > 90$ %, it can not be due to absorption by any amorphous phase present in the material. The quantum confinement effect may be excluded as the crystallite sizes are too large (> 3 nm) to induce this effect. Kaan Kalkan and Fonash, 1997 have proposed that crystallite size effects may alter phonon localization such that more phonons are available for indirect transitions, which would thus enhance absorption. The enhancement would thus be dependent on crystallite size and be larger for smaller crystallites as they experimentally verified. Vaněček et al, 1997 suggested that at least in the 1.2 -1.4 eV region, the enhanced absorption is merely an *apparent* absorption which is due to a longer effective optical path as a result of light scattering. The remaining *real* enhanced absorption coefficients above 1.4 eV are then due to high internal mechanical strain that can alter the **k**-vector selection rule and thus allow more optical transitions, and the enhanced absorption coefficients below 1.2 eV are attributed to disorder-induced broadening of the indirect absorption edge (Cody, 1992). Further, some measurement techniques detect a varying amount of subbandgap defect absorption, which is discussed in the paragraph on the defect density N_s. Although the origin for enhanced absorption above 1.0 eV is not exactly clear, the effect is favorable for applications in solar cells.

The photoconductivity σ_{ph} is also enhanced by the large absorption coefficients and optical scattering. In a coplanar structure, it might however appear to be limited due to grain boundaries perpendicular to the current path. Nevertheless, in spite of the indirect band gap, a value $> 1 \times 10^{-5}\ \Omega^{-1}\mathrm{cm}^{-1}$ should be achievable with 100 mW/cm^2, AM1.5 illumination, for a 1 μm film if it has appropriate surface roughness (200 nm peaks). The value of the dark conductivity is less ambiguous than that of the photoconductivity and for truly intrinsic poly-Si:H it is $\approx 1 \times 10^{-7}\ \Omega^{-1}\mathrm{cm}^{-1}$. A good photoresponse thus is $\sigma_{\mathrm{ph}}/\sigma_{\mathrm{d}} > 100$.

The dark conductivity activation energy E_{A} is strongly dependent on parts-per-million levels of impurities. Many laboratories find that their undoped microcrystalline silicon behaves like a n-type semiconductor with an activation energy around 0.4 eV, which raises the dark conductivity to $10^{-5}\ \Omega^{-1}\mathrm{cm}^{-1}$. This material is unsuitable for solar cells. The origin of this n-type behaviour is most likely the incorporation of oxygen, as this is known to act as an n-type dopant atom. This oxygen may originate from water vapor adsorbed on the walls of the deposition chamber or may be carried (as O_2 or as H_2O) with the feedstock gas. Even if it is carried with the feedstock gas there may not be a problem with the purity of the gas itself, but rather a result of adsorbants within gas lines, valves, or almost undetectable leakage of mass flow controllers. The latter origin of oxygen may be circumvented by using an oxygen getter for (re-)purifying the feedstock gas at the chamber gas inlet (Kroll et al., 1995). No purifier is needed if the deposition system is of ultrahigh vacuum (UHV) quality. Both Rath et al., 1997 and Wanka et al., 1997 report $E_{\mathrm{A}} = 0.54$ eV for a 1.1 eV band-gap material, which implies that the Fermi-level is at the center of the gap.

The hydrogen bonding. In hydrogenated microcrystalline and polycrystalline silicon, the Si-H stretching modes give detailed information that is correlated with the grain structure and probably as well with the electronic transport properties. Most hydrogen atoms are incorporated at grain boundaries or grain boundary tissue. Since they are bonded on the surface of crystallites facing intergrain voids the Si-H stretching modes give rise to absorption at around 2100 cm^{-1}. Usually a doublet is found at 2085 cm^{-1} and 2100 cm^{-1} (See Fig. 3.4), which can be correlated with the presence of both (220) and (111) orientations of Si crystallites. If a larger variety of crystal surfaces is present, Satoh and Hiraki, 1985 reported that as many as five peaks can be found ranging from 2085 cm^{-1} to 2155 cm^{-1}. In material with a relatively large fraction of amorphous grain boundary tissue (microcrystalline material as defined in Table 1.1) also bending modes can be found around 900 cm^{-1}. As can also be seen in Fig. 3.4, optimized polycrystalline thin films do not show any 2100 cm^{-1} absorption (Rath et al., 1997b), and instead have a small absorption only at 2000 cm^{-1}. This is interpreted as hydrogen bonded in very thin grain boundary regions facilitating a densely packed reconstruction of the Si network at adjacent grains.

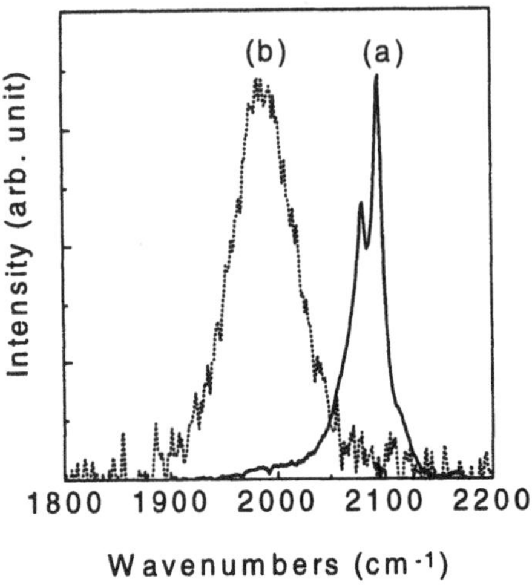

Figure 3.4 Infrared absorption spectra of polycrystalline silicon thin films. Curve (a) is for a film made using a H_2/SiH_4 ratio of 100 (high dilution), and (b) is for a film, made using a H_2/SiH_4 ratio of 10 (low dilution). For both cases the wire temperature, substrate temperature, and process pressure were 1800°C, 480°C, and 0.1 mbar, respectively.

The Hall mobility activation energy, the minority carrier diffusion length, and the majority carrier $\mu\tau$ -product. The Hall mobility for truly intrinsic poly-Si:H can not be easily determined because of the low carrier concentration which is between 10^{11} and 10^{12} cm^{-3}. The Hall mobility μ_H for good films currently is between 5 and 15 $cm^2V^{-1}s^{-1}$. The activation energy of μ_H gives important information on the carrier transport across grain boundaries. A low value for E_{μ_H} indicates that the potential barriers across the grain boundaries are small. Values as low as $E_{\mu_H} = 0.012$ eV have been measured.

The Steady-State Photocarrier Grating (SSPG) technique yields values for the ambipolar diffusion length L_D which is mainly determined by the $\mu\tau$ -product of the minority carriers. These values should again be treated with caution as they are limited by transport perpendicular to the grain boundaries. However, if they are high, then probably transport along the long axis of the grains will be adequate as well. For hot-wire deposited films, $L_D = 568$ nm has been reported (Rath et al., 1997b). This is 3 - 4 times higher than the value for device-quality amorphous silicon.

A good value for the majority carrier $\mu\tau$ -product is 1×10^{-6} cm^2/V using monochromatic red light. Values of 2×10^{-6} cm^2/V have been reported after post hydrogenation (Wanka et al., 1997) and 7×10^{-7} cm^2/V without any post-treatment (Rath et al., 1997).

The defect density N_s. Similar to the case of hydrogenated amorphous silicon, the midgap density of electronic defects depends on the characterization method used. For polycrystalline silicon, the differences between the results of various optoelectrical methods will even be larger as the current path may be along most of the grain boundaries in one case (such as in the CPM method) and perpendicular to them in another (such as in DBP in a sandwich structure). Optical methods (such as PDS) will be hard to interpret due to scattering at the top surface of the film due to native texture as well as at internal surfaces due to the grains. Nevertheless, PDS has been used for the determination of subbandgap absorption or defect absorption (Keppner et al., 1996) and may be useful for comparing equivalent samples. However, the role of the detected defects has not yet been addressed properly and a critical evaluation is necessary. For instance, while a significant subbandgap defect density can be determined for a poly-Si:H film from coplanar CPM, the same film as incorporated in a device shows complete absence of subbandgap absorption using DBP (Rath et al., 1997b). This is specifically the case in films with a strong anisotropy, where the DBP result is more relevant to cell operation.

Again, we suggest that ESR is the most straightforward technique for defect density determination. Two ESR centers have been observed in microcrystalline silicon. The resonances at $g \approx 2.005$ are attributed to the dangling bond centers at the grain boundaries and those that are isolated within the grains, and

Table 3.2. Criteria for 'device quality' polycrystalline silicon films.

	property	*requirement*
Minimum set	Crystalline fraction (Raman)	> 90 %
	Orientation of the grains (XRD)	predominantly (220)
	Band gap, indirect	1.0 - 1.1 eV
	AM1.5 100 mW/cm^2 photoconductivity	$> 1.5 \times 10^{-5}\ \Omega^{-1}\text{cm}^{-1}$
	Dark conductivity	$< 1.5 \times 10^{-7}\ \Omega^{-1}\text{cm}^{-1}$
	Activation energy	0.53 - 0.57 eV
Additional characteristics	Hall mobility activation energy	< 0.02 eV
	Minority carrier diffusion length (SSPG)	> 500 nm
	Mobility-lifetime product at 600 nm	$\geq 1 \times 10^{-7}$ cm^2/V
	Density of states (ESR method; g=2.005)	$\leq 1 \times 10^{17}$ cm^{-3} [a]
	Electron Spin Resonance (CESR line; g=1.998)	absent in intrinsic films

[a]This upper limit is only sufficiently low if other conditions are fulfilled as well.

another resonance at g = 1.998 exists in some samples. The latter resonance is called the CESR line and has been attributed to conduction electrons (CE) within the crystalline grains (Finger et al., 1994). Later, it was shown that the CESR signal clearly correlates with the n-type doping (Müller et al., 1997). Therefore, purely intrinsic poly-Si:H should not show the CESR line, as Rath et al., 1997b demonstrated in carefully prepared samples.

The defects detected at g $\approx$ 2.005, however, may not all be active as recombination centers if they are remote from the dominant current path. Therefore, it has been found that films with an ESR defect density of $\approx 10^{17}$ cm^{-3} can already be appropriate for use in solar cells. This value is much higher than that required for the i-layer in a-Si:H cells. Therefore, no absolute figure for the required ESR defect density can be given. It is of greater importance to establish that the activity of these centers in device structures is limited.

Summarizing, the criteria for device-quality intrinsic polycrystalline silicon thin films are collected in Table 3.2.

3.1.3 Hydrogenated amorphous silicon germanium alloys

The optimal design of a multijunction amorphous silicon solar cell uses the concept of spectrum splitting. In this concept the band gaps of the active layers of the component cells are selected such that the multijunction cell responds optimally to the solar spectrum.

Here, amorphous silicon technology has a distinct advantage as a large variety of compatible a-Si:H based materials with different band gaps can be produced by alloying it with other atoms such as Ge, C, O, or N. Hydrogenated amorphous silicon germanium (a-SiGe:H), silicon carbide (a-SiC:H), silicon oxide (a-SiO:H) or silicon nitride (a-SiN:H) alloys can easily be deposited by adding appropriate gases to the source gas mixture in the PECVD process.

In photovoltaic applications the purpose of alloying a-Si:H is to shift its optical absorption to higher or lower photon energies. Materials such as a-SiC:H, a-SiO:H, and a-SiN:H have wider band gaps than a-Si:H, and in a-SiGe:H alloys the absorption edge is shifted to lower photon energies. Due to the content of the solar spectrum, multibandgap solar cells gain more in performance if a-Si:H is combined with smaller band gap materials rather than with wider band gap materials. Narrow band gap materials are used in the bottom cell of tandem cells and in the middle and bottom component cells of triple junction solar cells.

The first a-SiGe:H alloys were prepared by Chevalier et al., 1977 by rf PECVD from a gas mixture of silane and germane. The Tauc optical band gap could easily be changed from 1.7 eV (a-Si:H) to 1.0 eV (a-Ge:H). However, it was also clear that the photoelectronic properties of a-SiGe:H deteriorated with increasing germanium content. This is explained by the existence of inhomogeneities both in the composition and structure of the films (Paul, 1988, Bauer et al., 1989). Using Raman spectroscopy, Bauer et al., 1989 demonstrated that a-$Si_{1-x}Ge_x$:H alloys with a Ge content up to x = 0.3 (corresponding to a Taux optical band gap E_g = 1.5 eV), exhibit a rigid Si-Si network and this material can be regarded as an extension of a-Si:H material. From this it can be understood that high quality a-$Si_{1-x}Ge_x$:H alloys were first obtained in the range of compositions x < 0.3. For a Ge content x > 0.3, where the a-$Si_{1-x}Ge_x$:H material behaves more like an alloy, the network structure is no longer dominated by a rigid Si-Si network but rather by the interaction between Si and Ge atoms. The traditional deposition parameters that have led to device-quality a-Si:H could not produce high quality a-SiGe:H material in this range of compositions. Improved a-SiGe:H was only achieved by high hydrogen dilution of the silane/germane mixture (Tanaka and Matsuda, 1986, Yang et al., 1991, Zeman et al., 1991) and/or by the use of fluorinated gas mixtures (Nozawa et al., 1983, Mackenzie et al., 1988).

One of the crucial parameters that determines the photovoltaic quality of amorphous semiconductors is the density distribution of the gap states. From subbandgap absorption measurements it was demonstrated that the slope of the valence band tail (the Urbach tail E_{V0}^{tail}) remains unchanged as the band gap is decreased to 1.25 eV. The values are close to that of device-quality a-Si:H at E_{V0}^{tail} = 50-60 meV (Guha, 1986). The measurements also showed that the density of defect states generally increases with increasing Ge content in the material. Since the Ge-Ge bond is weaker than the Si-Si bond one can expect that in disordered networks with identical topological structure, structures with a higher fraction of Ge exhibit a higher fraction of dangling bonds. Subbandgap absorption measurements can be combined with numerical simulations to determine the distribution of defect states in a-SiGe:H alloys (Carius et al., 1998). It was found that in a-$Si_{1-x}Ge_x$:H alloys in the range of $0 \leq x \leq 0.5$ the defect distributions are similar to those found in a-Si:H. Taking the position of the defect states as a reference level a decreased band gap can be interpreted as a shift of both band edges towards the defect states. From photoconductivity measurements it was shown that the conduction band tail broadens upon alloying of a-Si:H with Ge which in turn leads to a reduced electron drift mobility (Bauer et al., 1989).

The criteria for a device quality a-SiGe:H alloy with a Tauc optical gap around 1.45 eV for use in a tandem cell are listed in Table 3.3. The substrate temperature is an important deposition parameter that influences the growth and material properties of a-SiGe:H material. In order to obtain device-quality material as described in Table 3.3, the optimum deposition temperature for moderate hydrogen dilution ($\gamma = H_2/(SiH_4 + GeH_4) \approx 10$) is usually found at

230 to 280 °C (Kuznetsov et al., 1997, Lundszien et al., 1997). For a superstrate structure of a-Si:H solar cells this temperature is not suitable because of the possibility of thermal damage of the underlying layers and interfaces. The Sanyo group demonstrated that by using high hydrogen dilution (H_2/ SiH_4 > 27) it was possible to obtain a device-quality a-SiGe:H alloy with a cubic band gap of 1.32 eV at a substrate temperature of only 180 °C. Using this approach a stabilized efficiency of 9.5 % for an a-Si/a-SiGe superstrate-type submodule (area 1200 cm^2) has been achieved (Shima et al., 1998).

The use of only moderate hydrogen dilution already leads to a considerable decrease in the deposition rate, which is typically 1 Å/s for 1.45 eV a-SiGe:H material. For large scale production of a-Si:H based solar cells it is important to increase the deposition rate of a-SiGe:H materials. Replacing silane, which has about 3 times smaller dissociation rate than germane, with disilane, which has a dissociation rate similar to germane, is one approach. By increasing the discharge power at very high hydrogen dilution ($\gamma = H_2/\ (SiH_4 + GeH_4) \approx 80$) an increased deposition rate (above 2 Å/s) can be achieved without detrimental effects on the film microstructure or the density of states (Lundszien et al., 1997). The 70 MHz VHFCVD technique has been tested by Jones et al., 1998 as a high deposition rate (10 Å/s) process for the fabrication of a-SiGe:H alloy intrinsic layers. Canon uses low pressure microwave plasma CVD to achieve high deposition rate (40 Å/s) a-SiGe:H layers. With this technique a stabilized efficiency of 10 % was reported for a triple cell with an area of 0.25 cm^2 (Saito et al., 1996).

Table 3.3. Criteria for 'device quality' a-SiGe:H films with a Tauc band gap of 1.40 - 1.45 eV (cubic gap 1.31-1.32 eV).

	property	*requirement*
Minimum set	Dark conductivity	$< 5 \times 10^{-8}\ \Omega^{-1}cm^{-1}$
	AM1.5 100 mW/cm^2 photoconductivity	$> 1 \times 10^{-5}\ \Omega^{-1}cm^{-1}$
	Urbach energy	< 60 meV
	Activation energy	≈ 0.7 eV
Additional Characteristics	Density of dangling bond states	
	(Opto-)electrical methods (CPM, PDS)	$\leq 1 \times 10^{17}\ cm^{-3}$
	Microstructure parameter R^*	< 0.2
	Ge content	≈ 0.4
	H content	15 at.-%

3.2 DOPED MATERIALS

3.2.1 Doped amorphous silicon and silicon alloys

The performance of solar cells strongly depends on the properties of the p-layer. This layer should meet the conflicting requirements of high conductivity and low absorption. Unfortunately, boron tends to alloy with amorphous silicon rather than to substitutionally dope it, leading to a strong reduction in band gap. The band gap reduction is usually compensated by adding carbon to the lattice. Further, boron originating from diborane (B_2H_6) tends to cluster within the network which also degrades the transmission of the layer. Therefore, many investigations have been done to improve the quality of these layers by using monomeric dopant gases, such as $B(CH_3)_3$ (Sanyo), BF_3 (Fuji Electric), or $B(C_2H_5)_3$ (Tokyo Inst. of Technol.). All these dopant gases have been claimed to lead to superior p^+-layer properties. Of these gases, $B(CH_3)_3$ is at present actually used in production (De Neufville, 1997). It is thermally more stable than B_2H_6 and therefore p-layers can be made with better reproducibility, leading to better control of the properties (Shen et al., 1990).

Also p-type superlattices or multilayer structures have been proposed to optimize transmission and conductivity in a more controlled fashion. For instance, $(\text{a-SiC}_x/\text{a-SiC}_y)_n$ or $(\text{a-Si:H/a-C:H})_n$ multilayers have been reported by Ashida, 1993 to possess a five times higher dark conductivity than that of a conventional p-layer while maintaining the same low absorption coefficient. More of interest to manufacturing, improved p^+-layers ($E_g = 2.15$ eV and $\sigma_{\text{ph}} = 1.2 \times 10^{-6}\ \Omega^{-1}\text{cm}^{-1}$), as well as improved buffer layers (Ichikawa et al., 1992) have been reported by using alloying gases other than CH_4, such as CO_2, which yields an a-SiO:H alloy. A high efficiency has been obtained by using an "alternately repeating deposition and hydrogen plasma treatment method" (ADHT; also known as "chemical annealing") for preparing improved buffer layers (Tanaka et al, 1993).

Typical criteria for doped a-SiC:H layers for application in p^+-i-n^+ or n^+-i-p^+ solar cells are listed in Table 3.4. The desired properties of p-type a-SiC:H layers are typically obtained at a SiH_4:CH_4:B_2H_6 flow ratio of 1:2:0.001, where H_2 can be added to further optimize properties and uniformity. The desired properties of n-type layers are relatively easily obtained from mixtures of SiH_4 and PH_3 in a ratio of 1:0.0025, where H_2 can again be used to enhance the performance depending on the reactor geometry.

3.2.2 Doped microcrystalline silicon

Doped microcrystalline silicon layers are used in solar cells at the tunnel-recombination junction of multijunction cells and as window layers providing a good contact with transparent or metal electrodes. They are also applied as the emitter layers in c-Si solar cells with a deposited heterojunction, in bipolar transistors, thin film transistors, and in light emitting diodes. In solar cells, the much higher doping efficiency of microcrystalline silicon as compared to amorphous silicon leads to an enhanced V_{oc} and FF, due to the improved built-in potential difference over the i-layer. Furthermore, the relatively low absorption coefficient decreases parasitic absorption in the doped layers and thus leads to higher J_{sc} values. The first successful application of microcrystalline doped layers was probably demonstrated by Guha et al., 1986.

Using PECVD, these layers are usually prepared from highly hydrogen diluted silane ($H_2/SiH_4 > 100$), at high power densities. A high substrate temperature during deposition results in a higher crystalline volume fraction, but often such temperatures are not compatible with previously deposited layers. A high process pressure leads to larger grains but does delay the nucleation as was recently found (Rath and Schropp, 1998).

Table 3.4. Criteria for doped a-Si:H layers for application in solar cells.

property	*requirement*	
	p-type a-Si:H	*n-type a-Si:H*
Conductivity (Ω^{-1}cm^{-1})	$> 10^{-5}$	$> 10^{-3}$
Conductivity for a 20 nm thick film (Ω^{-1}cm^{-1})	$> 10^{-7}$	$> 10^{-4}$
Band gap, Tauc (eV)	> 2.0	> 1.75
Activation energy (eV)	< 0.5	< 0.3
Absorption coefficient at 600 nm (cm^{-1})	$\leq 1 \times 10^4$	$\leq 3 \times 10^4$
at 400 nm (cm^{-1})	$\leq 3 \times 10^5$	

Table 3.5. Properties of microcrystalline doped films obtained with 13.56 MHz PECVD.

Type	*p-type μc-Si:H* [a]		*n-type μc-Si:H* [b]	
Thickness	*300 nm*	*20 nm*	*70 nm*	*13 nm*
Conductivity (Ω^{-1}cm^{-1})	1.5	2.6×10^{-2}	20	2.5
Band gap, Tauc (eV)	1.4	-	1.9	-
Activation energy (eV)	0.025	0.059	0.01	0.029
Crystalline volume fraction (%)	67	17		

[a]Rath et al., 1995, Rath and Schropp, 1998; [b]Landweer et al., 1994.

In order for the solar cell to benefit from the special properties of microcrystalline layers in doped window layers and tunnel-recombination junction layers, the microcrystalline volume fraction should be above the lower limit for percolation conduction (V_f = 16 % for spherical microcrystallites; Tsu et al., 1982) within the first 20 nm of deposition. This can be verified by spectroscopic ellipsometry or Raman spectroscopy (as discussed in Section 3.1.2. The peak position due to the crystalline component of very thin microcrystalline layers is shifted to lower wavenumbers from the pure crystalline peak position (eg. 514 cm^{-1} instead of 520 cm^{-1}) due to the grain size effect (Iqbal and Vepřek, 1982). A thick film, i.e., ≈ 1 μm, can easily reach a conductivity of 10 $\Omega^{-1}cm^{-1}$ if it is n-type and 1 $\Omega^{-1}cm^{-1}$ if it is p-type. However, the achievement of thin layers that have a large crystalline volume fraction is difficult (Tsai et al., 1990) and also critically dependent on the nature of previously deposited layers. For n-type layers, the conductivity of 20 nm thick layers is orders of magnitude lower, whereas p-type layers may appear to be completely amorphous even though they are grown under "microcrystalline" conditions (Goldstein et al., 1988). The difficulty specifically in obtaining thin p-layers has been associated with the presence of boron that effectively hampers crystalline growth (Prasad et al., 1991). Several methods have been proposed to help nucleation such that also thin microcrystalline layers could be formed. At an early stage in the development of PECVD, the use of fluorinated gases was proposed by Guha et al., 1986. The use of fluorine-containing gases still plays a role at United Solar in obtaining good quality microcrystalline doped layers. Actually, the earliest report on fluorinated "amorphous" Si-H alloys by Madan et al, 1979 with conductivities in excess of 5 $\Omega^{-1}cm^{-1}$, probably not knowingly dealt with microcrystalline material. Recently, the layer-by-layer technique has been used to enhance crystallinity in a thin layer (Miyamoto at al., 1997). The layer-by-layer technique can overcome the incubation phase that has been observed when depositing thin microcrystalline layers (Roca i Cabarocas et al., 1995). An excitation frequency that is higher in the VHF range promotes crystallinity (Oda et al., 1988), most likely due to better dissociation of H_2 into atomic hydrogen that etches silicon from strained bonds, and the reduction of ion bombardment energy (Kondo et al., 1996). Thin doped microcrystalline layers can also be obtained by conventional PECVD at 13.56 MHz by adapting the parameter regime specifically for thin layers, rather than merely decreasing the deposition time. A process pressure that is reduced with respect to the optimum pressure for thick layers yields a smaller crystalline volume fraction but an earlier onset of crystallite nucleation (Rath and Schropp, 1998). Moreover, in contrast to Goldstein et al., 1988 a 20 nm p-type μc-Si:H layer could be formed directly on SnO_2:F-coated glass, without noticeable deterioration of the SnO_2:F due to the hydrogen plasma. Typical materials properties obtained with the conventional 13.56 MHz PECVD technique are listed in Table 3.5. Other methods than conventional PECVD such as ECR-CVD (Hattori et al., 1987) and photo-CVD (Konagai et al., 1985) have been quite successful in producing microcrystalline doped layers that are thin enough for use in solar cells. In particular, Hattori et al., 1987 achieved a V_{oc} of 0.97 V due to the high optical band gap and low activation energy of thin p-type μc-Si:H prepared by ECR-CVD.

References

Bauer, G.H., C.E. Nebel, M.B. Schubert, and G. Schumm, *Band tailing and transport in a-SiGe:H alloys*, in: Amorphous Silicon Technology - 1989, edited by A. Madan, M.J. Thompson, P.C. Taylor, Y. Hamakawa, and P.G. LeComber, Materials Research Society Symp. Proc. **149** (1989) 485-496.

Beck, N., J. Meier, J. Fric, Z. Remes, A. Poruba, R. Flückiger, J. Pohl, A. Shah, and M. Vaněček, *Enhanced optical absorption in microcrystalline silicon*, J. Non-Cryst. Solids **198-200** (1996) 903-906.

Beyer, W., and M.S. Abo Ghazala, *Absorption strengths of Si-H vibrational modes in hydrogenated silicon*, in: Amorphous and Microcrystalline Silicon Technology - 1998, edited by R. Schropp, H. Branz, S. Wagner, M. Hack, and I. Shimizu, Materials Research Society Symp. Proc. **507** (1998) in print.

Brendel, R., M. Hirsch, M. Stemmer, U. Rau, and J.H. Werner, *Internal quantum efficiency of thin epitaxial silicon solar cells*, Appl. Phys. Lett. **66** (1995) 1261-1263.

Brodsky, M.H., M. Cardona, and J.J. Cuomo, *Infrared and Raman spectra of the silicon-hydrogen bonds in amorphous silicon prepared by glow discharge and sputtering*, Phys. Rev. B **16** (1977) 3556-3571.

Bustarret, E., M.A. Hachinia and M. Brunel, *Experimental determination of the nanocrystalline volume fraction in silicon thin films from Raman spectroscopy*, Appl. Phys. Lett. **52** (1988) 1675-1677.

Carius, R., H. Stiebig, F. Siebke, and J. Fölsch, *Defect distribution in a-SiGe:H*, in: Amorphous and Microcrystalline Silicon Technology - 1998, edited by R. Schropp, H. Branz, S. Wagner, M. Hack, and I. Shimizu, Materials Research Society Symp. Proc. **507** (1998) in print.

Chevallier, J., H. Wieder, A. Onton and C.R. Guarnieri, *Optical properties of amorphous $Si_xGe_{1-x}(H)$ alloys prepared by r.f. glow discharge*, Solid State Comm. **24** (1977) 867-869.

Cody, G.D., *Urbach edge of crystalline and amorphous silicon: a personal review*, J. Non-Cryst. Solids **141** (1992) 3-15.

Daey Ouwens, J., and R.E.I. Schropp, *Hydrogen microstructure in hydrogenated amorphous silicon*, Phys. Rev. B **54** (1996) 17759-17562.

Den Boer, W., *Determination of the midgap density of states in a-Si:H using space-charge-limited current measurements*, J. de Phys. C4 **42** (1981) 451-454.

De Neufville, J.P., private communication, 1997.

Fauchet, P.M., and I.H. Campbell, *Raman spectroscopy of low-dimensional semiconductors*, Crit. Rev. Sol. State Mat. Sci. **14** (1988) S79-S101.

Finger, F., C. Malten, P. Hapke, R. Carius, R. Flückiger, and H. Wagner, *Free electrons and defects in microcrystalline silicon studied by electron spin resonance*, Phil. Mag. Lett. **70** (1994) 247-254.

Fritzsche, H., and M. Tanielian, *Problems regarding the conductance in a-Si:H films*, in: Conference on Tetrahedrally Bonded Amorphous Semiconductors, Carefree, Arizona, 1981, edited by R.A. Street, D.K. Biegelsen, and J.C. Knights (American Institute of Physics, 1981) AIP Conference Proceedings **73** (1981) 318-322.

Goldstein, B., C.R. Dickson, I.H. Campbell, and P.M. Fauchet, *Properties of p^+ microcrystalline films of a-SiC:H deposited by conventional rf glow discharge*, Appl. Phys. Lett. **53** (1988) 2672-2674.

Green, M.A., and M.J. Keevers, *Optical properties of intrinsic silicon at 300 K*, Progress in Photovoltaics: Research and Applications **3** (1995) 189-192.

Guha, S., *Materials considerations for high efficiency and stable amorphous silicon alloy solar cells*, in: Proc. of the Int. Workshop on Amorphous Semiconductors, eds. H. Fritsche, D.X. Han, and C.C. Tsai, (1987 World Scientific Publishing Co.) 313-320.

Guha, S., J. Yang, P. Nath, and M. Hack, *Enhancement of open circuit voltage in high efficiency amorphous silicon alloy solar cells*, Appl. Phys. Lett. **49** (1986a) 218-219.

Hattori, Y., D. Kruangam, K. Katoh, Y. Nitta, H. Okamoto and Y. Hamakawa, *High-conductivity wide band gap p-type a-SiC:H prepared by ECR-CVD and its application to high efficiency a-Si basis solar cells*, in: Proc. of the 19th IEEE PV Specialists Conf., 1987, 689-694.

Iqbal, Z., S. Vepřek, A.P. Webb, and P. Capezutto, *Raman scattering from small particle size polycrystalline silicon*, Solid State Commun. **37** (1981) 993-996.

Iqbal, Z., and S. Vepřek, *Raman scattering from hydrogenated microcrystalline and amorphous silicon*, J. Phys. C: Solid State Phys. **15** (1982) 377-392.

Jackson, W.B., N.M. Amer, A.C. Boccara, and D. Fournier, *Photothermal Deflection Spectroscopy and detection*, Appl. Optics **20** (1981) 1333-1334.

Jackson, W.B., S.M. Kelso, C.C. Tsai, J.W. Allen, S.-J. Oh, *Energy dependence of the optical matrix element hydrogenated amorphous and crystalline silicon*, Phys. Rev. B **31** (1985) 5187-5198.

Jones, S.J., X. Deng, T. Liu and M. Izu, *Preparation of a-Si:H and a-SiGe:H i-layers for nip solar cells at high deposition rates using a very high frequency technique*, in: Amorphous and Microcrystalline Silicon Technology - 1998, edited by R. Schropp, H. Branz, S. Wagner, M. Hack, and I. Shimizu, Materials Research Society Symp. Proc. **507** (1998) in print.

Kaan Kalkan, A., and S.J. Fonash, *Control of enhanced optical absorption in μc-Si*, in: Amorphous and Microcrystalline Silicon Technology - 1997, edited by S. Wagner, M. Hack, E.A. Schiff, R. Schropp, and I. Shimizu, Materials Research Society Symp. Proc. **467** (1997) 319-324.

Keppner, H., P. Torres, J. Meier, R. Platz, D. Fischer, U. Kroll, S. Dubail, J.A. Anna Selvan, N. Pellaton Vaucher, Y. Ziegler, R. Tscharner, Ch. Hof, N. Beck, M. Goetz, P. Pernet, M. Goerlitzer, N. Wyrsch, J. Veuille, J. Cuperus, A. Shah, and J. Pohl, *The "Micromorph" cell: a new way to high-efficiency low-temperature crystalline silicon thin-film cell manufacturing?*, in: Advances in Microcrystalline and Nanocrystalline Semiconductors - 1996, edited by R.W. Collins, P.M. Faucher, I. Shimizu, J.C. Vial, T. Shimada, and A.P. Alivisatos, Materials Research Society Symp. Proc. **452** (1996) 865-876.

Klazes, R.H., M.H.L.M. van den Broek, J. Bezemer, and S. Radelaar, *Determination of the optical bandgap of amorphous silicon*, Phil. Mag. B **45** (1982) 377-383.

Klug, H.P., and L.E. Alexander, *X-ray diffraction procedure*, (John Wiley & Sons, New York, 1974).

Konagai, M., H. Takai, W.Y. Kim, and K. Takahasi, *Preparation of amorphous silicon and related semiconductors by photochemical vapor deposition and its application to solar cells*, in: Proc. of the 18th IEEE PV Specialists Conf., 1985, 1372-1377.

Kondo, M., Y. Toyoshima, A. Matsuda, and K. Ikuta, *Substrate dependence of initial growth of microcrystalline silicon in plasma-enhanced chemical vapor deposition*, J. Appl. Phys. **80** (1996) 6061-6063.

Kroll, U., J. Meier, H. Keppner, S.D. Littlewood, I.E. Kelly, and A. Shah, and P. Giannoulès, *Origin and incorporation mechanism for oxygen contaminants in a-Si:H and μc-Si:H films prepared by the very high frequency (70 MHz) glow discharge technique*, in: Amorphous Silicon Technology - 1995, edited by M. Hack, E.A. Schiff, A. Madan, M. Powell, and A. Matsuda, Materials Research Society Symp. Proc. **377** (1995) 39-44.

Kumeda, M., and T. Shimizu, *ESR in hydrogenated amorphous silicon*, Jap. J. Appl. Phys. **19** (1980) L197-L200.

Kuznetsov, V.I., M. Zeman, L.L.A. Vosteen, B.S. Girwar, and J.W. Metselaar, *Electrical and optical properties of plasma-deposited a-SiGe:H alloys: Role of growth temperature and post-growth anneal*, J. Appl. Phys. **80** (1996) 6496-6504.

Landweer, G.E.N., C.H.M. van der Werf, R.W. Stok, J.W. Metselaar, and R.E.I. Schropp, *The application of micro-crystalline n-type Si in amorphous silicon tandem solar cells*, 12th International E.C. Photovoltaic Solar Energy Conference 1994, Eds. R. Hill, W. Palz, and P. Helm, (H.S. Stephens and Associates, 1994) 1284-1287

Lang, D.V., J.D. Cohen, and J.P. Harbison, *Measurement of the density of gap states in hydrogenated amorphous silicon by space charge spectroscopy*, Phys. Rev. B **25** (1982) 5285-5320.

Langford, A.A., M.L. Fleet, B.P. Nelson, W.A. Lanford, and N. Maley, *Infrared absorption strength and hydrogen content of hydrogenated amorphous silicon*, Phys. Rev. B **45** (1992) 13367-13377.

Lundszien, D., J. Fölsch, F. Finger and H. Wagner, *Is there an optimization limit for hydrogenated amorphous silicon-germanium for solar cell applications?*, Proceedings of the 14th European Photovoltaic Solar Energy Conference, June 30 - July 4, Barcelona, Spain, Eds. H.A. Ossenbrink, P. Helm, and H. Ehmann (H.S. Stephens and Associates, 1997) 578-581.

Mackenzie, K.D., P.G. LeComber, and W.E. Spear, *The density of states in amorphous silicon determined by space-charge-limited current measurements*, Phil. Mag. B **46** (1982) 377-389.

Mackenzie, K.D., J.H. Burnett, J.R. Eggert, Y.M. Li, and W. Paul, *Comparison of the structural, electrical, and optical properties of amorphous silicon-germanium alloys produced from hydrides and fluorides*, Phys. Rev. B **38** (1988) 6120-6136.

Madan, A., S.R. Ovshinksy, and E. Benn, *Electrical and optical properties of amorphous Si:F:H alloys*, Phil. Mag. B **40** (1979) 259-277.

Mahan, A.H., J. Carapella, B.P. Nelson, R.S. Crandall, and I. Balberg, *Deposition of device quality, low H content amorphous silicon*, J. Appl. Phys. **69** (1991) 6728-6730.

Maley, N., *Critical investigation of the infrared transmission data analysis of hydrogenated amorphous silicon alloys*, Phys. Rev. B **4** (1992) 2078-2085.

Matsuda, A., *Formation kinetics and control of microcrystallite in μc-Si:H from glow discharge plasma*, J. Non-Cryst. Solids **59 & 60** (1983) 767-774.

Meyer, W., and H. Neldel, *Über die Beziehungen zwischen die Energiekonstanten Epsilon und der Mengenkonstanten a in der Leitwerts-Temperaturformel bei oxydischen Halbleitern*, Z. Techn. Phys. **18** (1937) 588-593.

Miyamoto, Y., A. Miida, and I. Shimizu, in: Advances in Microcrystalline and Nanocrystalline Semiconductors - 1996, edited by R.W. Collins, P.M. Faucher, I. Shimizu, J.C. Vial, T. Shimada, and A.P. Alivisatos, Materials Research Society Symp. Proc. **452** (1997) 995.

Moddel, G., D.A. Anderson, and W. Paul, *Derivation of the low-energy optical-absorption spectra of a-Si:H from photoconductivity*, Phys. Rev. B **22** (1980) 1918-1925.

Müller, J., F. Finger, C. Malten, and H. Wagner, *Photocarrier recombination in microcrystalline silicon studied by light induced electron spin resonance transients*, in: Advances in Microcrystalline and Nanocrystalline Semiconductors - 1996, edited by R.W. Collins, P.M. Faucher, I. Shimizu, J.C. Vial, T. Shimada, and A.P. Alivisatos, Materials Research Society Symp. Proc. **452** (1997) 827-832.

Nozawa, K., Y. Yamaguchi, J. Hanna, and I. Shimizu, *Preparation of photoconductive a-SiGe alloy by glow discharge*, J. Non-Cryst. Solids **59 & 60** (1983) 533-536.

Oda, S., J. Noda, and M. Matsumura, *Preparation of a-Si:H films by VHF plasma CVD*, in: Amorphous Silicon Technology, edited by A. Madan, M.J. Thompson, P.C. Taylor, P.G. LeComber, and Y. Hamakawa, Materials Research Society Symp. Proc. **118** (1988) 117-122.

Paul, W., *Heterogeneities in a-SiGe:H alloys*, in: Amorphous Silicon and Related Materials, ed. H. Fritsche, (World Scientific Publishing Company, 1988) 63-79.

Prasad, K., U. Kroll, F. Finger, A. Shah, J.-L. Dorier, A. Howling J. Baumann, and M. Schubert, in: Amorphous Silicon Technology - 1991, edited by A. Madan, Y. Hamakawa, M.J. Thompson, P.C. Taylor, and P.G. LeComber, Materials Research Society Symp. Proc. **219** (1991) 383.

Rath, J.K., M.B. von der Linden, E.H.C. Ullersma, and R.E.I. Schropp, *Achievement of thin (15 nm) p-type microcrystalline silicon films by using an adapted deposition parameter regime*, Proceedings of the 13th EC Photovoltaic Solar Energy Conference, edited by W. Freiesleben, W. Palz, H.A. Ossenbrink, and P. Helm (H.S. Stephens and Ass., Bedford, UK, 1995) 1712-1715.

Rath, J.K., A. Barbon, and R.E.I. Schropp, *Limited influence of grain boundary defects in hot-wire CVD polysilicon films on solar cell performance*, J. Non-Cryst. Solids **227-230** (1998) 1277-1281.

Rath, J.K., H. Meiling and R.E.I. Schropp, *Purely intrinsic poly-Si films for n-i-p solar cells*, Jpn. J. Appl. Phys. **36** (1997) 5436-5443.

Rath, J.K., and R.E.I. Schropp, *Incorporation of p-type microcrystalline silicon films in amorphous silicon based solar cells in a superstrate structure*, Solar Energy Materials and Solar Cells **53** (1998) 189-203.

Ritter, D., E. Zeldov, and K. Weiser, *Steady-state photocarrier grating technique for diffusion length measurements in photoconductive insulators*, Appl. Phys. Lett. **49** (1986) 791-793.

Roca i Cabarocas, P., N. Layadi, T. Heitz, B. Drévillon, and I. Solomon, *Substrate selectivity in the formation of microcrystalline silicon: mechanisms and technological consequences*, Appl. Phys. Lett. **66** (1995) 3609-3611.

Saito, K., M. Sano, J. Matsuyama, M. Higasikawa, K. Ogawa and I. Kajita, *The light Induced Degradation of the a-Si:H Alloy Cell Deposited by the Low Pressure Microwave PCVD at High Deposition Rate*, Technical Digest of the International PVSEC-9, Miyazaki, Japan, (1996) 579.

Satoh, T., and A. Hiraki, *Detailed study of Si-H stretching modes in μc-Si:H film through second derivative IR spectra* Jpn. J. Appl. Phys. **24** (1985) L491-L494.

Shen, D., H. Chatham, and R.E.I. Schropp, *$B(CH_3)_3$ as p layer doping gas*, in: Amorphous Silicon Technology - 1990, edited by P.C. Taylor, M.J. Thompson, P.G. LeComber, Y. Hamakawa, and A. Madan, Materials Research Society Symp. Proc. **192** (1990) 523-528.

Shima, M., M. Isomura, E. Maruyama, S. Okamoto, H. Haku, K. Wakisaka, S. Kiyama and S. Tsuda, *Development of stable a-Si/a-SiGe tandem solar cell submodules deposited by a very high hydrogen dilution at low temperature*, in: Amorphous and Microcrystalline Silicon Technology - 1998, edited by R. Schropp, H. Branz, S. Wagner, M. Hack, and I. Shimizu, Materials Research Society Symp. Proc. **507** (1998) in print.

Stutzmann, M., D.K. Biegelsen, and R.A. Street, *Detailed investigation of doping in hydrogenated amorphous silicon and germanium*, Phys. Rev. B **35** (1987) 5666-5701.

Tanaka, K., and A. Matsuda, *Deposition kinetics and structural control of highly photosensitive a-SiGe:H alloys*, in: Materials Issues in Amorphous Semiconductor Technology, edited by D. Adler, Y. Hamakawa, and A. Madan, Materials Research Society Symp. Proc. **70** (1986) 245-255.

Tauc, J., in: *Optical properties of solids*, edited by F. Abeles (North-Holland Publ. Co., 1972) 277-313.

Tsai, C.C., G.B. Anderson, and R. Thompson, *Growth of amorphous, microcrystalline, and epitaxial silicon in low temperature plasma deposition*, in: Amorphous Silicon Technology - 1990, edited by P.C. Taylor, M.J. Thompson, P.G. LeComber, Y. Hamakawa, and A. Madan, Materials Research Society Symp. Proc. **192** (1990) 475-480.

Tsu, R., J. Gonzalez-Hernandez, S.S. Chao, S.C. Lee, and K. Tanaka, *Critical volume fraction of crystallinity for conductivity percolation in phosphorus-doped Si:F:H alloys*, Appl. Phys. Lett. **40** (1982) 534-535.

Tzolov, M.B., N.V. Tzenov, D.I. Dimova-Malinovska, *Analysis of the infrared transmission data of amorphous silicon and amorphous silicon alloy films*, J. Phys. D: Appl. Phys. **26** (1993) 111-118.

Vaněček, M., A. Poruba, Z. Remeš, N. Beck, and M. Nesládek, *Optical properties of microcrystalline materials*, J. Non-Cryst. Solids **227-230** (1998) 967-972.

Vepřek, S., F.-A. Sarott, and Z. Iqbal, *Effect of grain boundaries on the Raman spectra, optical absorption, and elastic light scattering in nanometer-sized crystalline silicon*, Phys. Rev. B **36** (1987) 3344-3350.

Wanka, H.N., M.B. Schubert, A. Hierzenberger, V. Baumung, *Prospects of microcrystalline silicon from hot-wire CVD for photovoltaic applications*, Proceedings of the 14th European Photovoltaic Solar Energy Conference, June 30 - July 4, Barcelona, Spain, Eds. H.A. Ossenbrink, P. Helm, and H. Ehmann (H.S. Stephens and Associates, 1997) 1003-1006.

Ward, L., *The optical constants of bulk materials and films*, (IOP Publishers, Bristol, 1994, 2nd ed.) 246.

Weber, K.J., A. Stevens, and A.W. Blakers, *Investigation of the effect of various process parameters during liquid phase epitaxy of silicon on silicon substrates*, Proceedings of the 13th EC Photovoltaic Solar Energy Conference, edited by W. Freiesleben, W. Palz, H.A. Ossenbrink, and P. Helm, (H.S. Stephens and Ass., Bedford, U.K., 1995) 1590-1593.

Yang, L., L. Chen, and A. Catalano , *The effect of hydrogen dilution on the deposition of SiGe alloys and the device stability*, in: Amorphous Silicon Technology - 1991, edited by A. Madan, Y. Hamakawa, M.J. Thompson, P.C. Taylor, and P.G. LeComber, Materials Research Society Symp. Proc. **219** (1991) 259-264.

Zanzucchi, P.J., C.R. Wronski, and D.E. Carlson, *Optical and photoconductive propoerties of discharge-produced amorphous silicon*, J. Appl. Phys. **48** (1977) 5227-5236.

Zeman, M., I. Ferreira, M.J. Geerts, and J.W. Metselaar, *The effect of hydrogen dilution on glow discharge a-SiGe:H alloys*, Solar Energy Materials **21** (1991) 255 - 265.

Zhu, M., and H. Fritzsche, *Density of states and mobility-lifetime product in hydrogenated amorphous silicon, from thermostimulated conductivity and photoconductivity measurements*, Phil. Mag. B **53** (1986) 41-54.

Ziman, J.M., *Models of disorder*, (Cambridge University Press, 1979).

4 TECHNOLOGY OF SOLAR CELLS

We chose to examine discharge-produced a-Si as a potential solar cell material because of its unusual electrical and optical properties.

—D.E. Carlson and C.R. Wronski, 1976

4.1 PRINCIPLES OF SOLAR CELL OPERATION

The operation principle of a solar cell is based on a sufficiently long lifetime of photoexcited electrons and holes such that they can become spatially separated and thus contribute to the net current. A "classical" solar cell consists of a p-type and an n-type domain formed within the same material, thus creating a p-n junction. Geminate recombination (i.e. recombination of photoexcited electrons and holes associated with each other by one photon absorption event) is prevented in such a cell by the internal field existing at the p-n junction. After separation, carrier recombination may still take place. In fact, non-geminate recombination is the most important loss mechanism in a solar cell.

In amorphous and microcrystalline semiconductors, a solar cell can not be constructed by merely stacking a p-type and an n-type thin film. Such a device would not show photovoltaic actvity, as the lifetime of photoexcited free charge carriers is too short to separate any significant fraction of them. The short lifetime is the result of the high concentration of defects associated with doping (either p- or n-type) as a result from autocompensation (Street, 1991). An undoped layer with a low defect density must thus be incorporated between the p- and n-type layer such that the mean free path of the slowest carriers (holes) is large enough to spatially separate them from the photogenerated electrons in this region and to collect them (Carlson and Wronski, 1976). The mean free path in the presence of a field is determined by the mobility-lifetime product $\mu\tau$ (See Section 3.1.1), where μ is the mobility and τ is the lifetime. In device-quality a-Si:H, $\mu\tau = 10^{-8}$ cm^{-2}/V for holes and $\mu\tau = 10^{-6}$ cm^{-2}/V for electrons. Even in a very weak field, the drift length $\ell = \mu\tau E$ for holes is still large enough to collect most of them. However, the diffusion length is only 100 - 200 nm, and therefore the presence of a field-free region of comparable length at the center of the intrinsic layer would drastically reduce the collection efficiency.

The built-in voltage V_{bi} between the highly doped (p^+ and n^+) layers is distributed non-uniformly over the intrinsic layer (i-layer). This is due to the spatially and energetically continuous density of states within the band gap of the i-layer (see also Chapter 6). As the density of states is distributed over the band gap with increasing concentration toward the valence and conduction band edges, the regions near the p^+ and n^+-layers contain a relatively high space charge density compared to the middle part of the i-layer. Poisson's equation (Equation 6.1)) thus dictates that the electric field is the highest ($\approx 10^5$ V/cm) near the p^+ and the n^+ contacts (Hack and Shur, 1985), whereas the total V_{bi} is still large enough for thin cells to maintain a low field ($\approx 10^4$ V/cm) in the bulk of the i-layer. In crystalline Si terminology, the i-layer is completely depleted. Fig. 4.1 illustrates the potential distribution in a p^+-i-n^+ solar cell as calculated using the defect pool model (Section 6.5.3). Under normal operating conditions, a low-field region is clearly present. Further illustrations resulting from device modeling can be found in Section 8.2.

A local collapse of the field can occur due to light-induced defects and the associated spatial redistribution of charge. As these defects, in addition, reduce the $\mu\tau$ for holes (via their lifetime τ), light-induced defect creation is quite effective in reducing the efficiency. The light-induced changes in bulk materials and solar cells will be discussed in Chapter 5.

In amorphous silicon p^+-i-n^+ solar cells, as well as in equivalent microcrystalline cells, almost all photocarriers originate from a region within the cell where an electric field is present. However, when the cell is operated near the maximum power point, (i.e. under forward bias) the field is reduced such that a virtually field-free region develops centrally in the i-layer. The total deple-

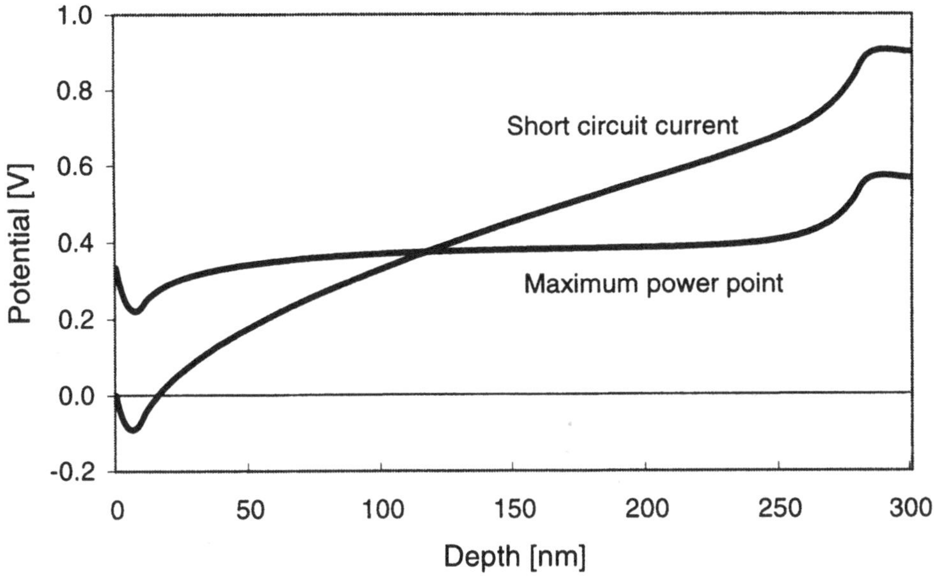

Figure 4.1. Potential distribution in a p^+-i-n^+ solar cell calculated using the defect pool model, under short circuit conditions and at the maximum power point.

tion region then becomes smaller than the thickness of the i-layer and thus the photocurrent is a function of the forward voltage. Therefore, these p^+-i-n^+ structures are typically called *drift* type solar cells, in contrast to, e.g., single crystal solar cell designs where the photocurrent is independent of external voltage (*diffusion*-type devices). Due to the enhanced curvature of the illuminated I-V curve with respect to the I-V curve in the dark, the maximum obtainable fill factor is lower in a drift type device than in a diffusion type device. In a typical drift type p^+-i-n^+ device, at a certain forward voltage beyond the open circuit voltage, the photocurrent will change sign, which is apparent experimentally from the intersection of the I-V curves measured under illumination and in the dark.

4.2 SUPERSTRATE SOLAR CELLS

In superstrate-type solar cells, the carrier on which the various thin film materials are deposited serves as a window to the cell. Usually in superstrate cells, glass is used as the carrier. A more technologically challenging version of a superstrate configuration is a cell on a transparent thin polymer foil. This type of cell is bound to have performance limitations as the limited tempera-

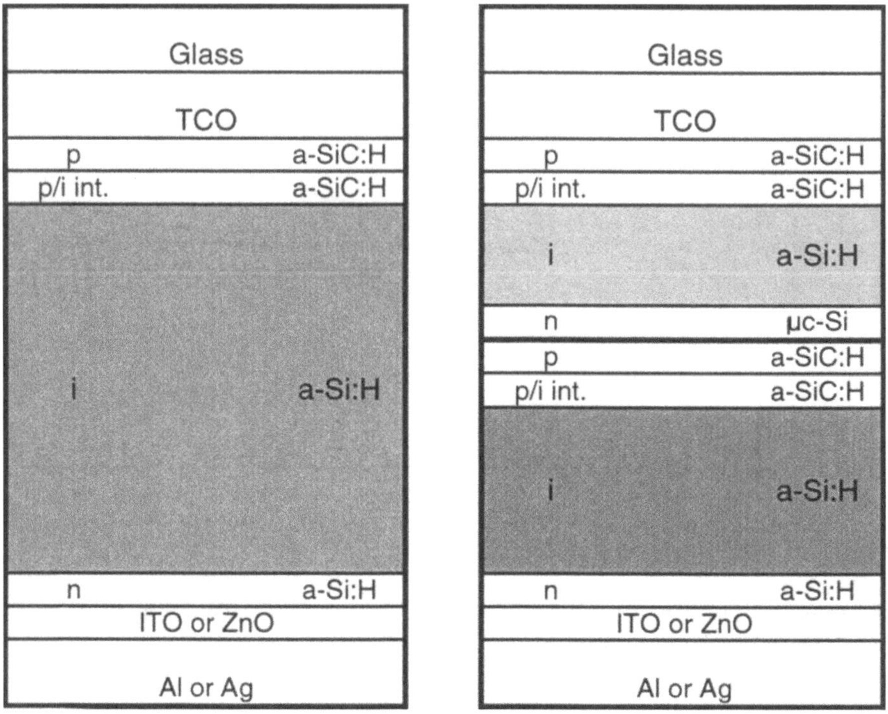

Figure 4.2. Cross-sectional diagram of a superstrate single junction and tandem solar cell.

ture resistance of most transparent plastics dictates the maximum temperature allowable in further process steps. Therefore, the present world production volume of superstrate solar cells consists primarily of single or multijunction p^+-i-n^+ structures on glass. A cross section of superstrate solar cells is shown in Fig. 4.2.

4.2.1 Front electrode technology

In superstrate devices, the front electrode usually consists of a transparent conductive oxide (TCO) layer such as SnO_2:F, doped ZnO, or In_2O_3:Sn (ITO). The requirements for the electrode material are (i) a high transmission (preferably $> 85\ \%$) in the entire wavelength region where the photovoltaic absorber layers are active. In multibandgap multijunction cells the active region extends into the near infrared spectrum, (ii) a low sheet resistance (preferably $< 10\ \Omega/\Box$; for multijunction cells $< 20\ \Omega/\Box$) to minimize series resistance losses, (iii) a low contact resistance with the p-type layer of the p^+-i-n^+ structure, (iv) a rough surface to increase the optical path of the light by scattering so that the absorption in the active layers is enhanced, while the morphology should be such that shunting paths, pinholes, or local depletion is avoided. This is usually achieved by utilizing the native surface texture of polycrystalline metal oxide thin films, (v) chemical resistance against the strongly reducing H ambient during chemical vapor deposition.

The ability of a TCO layer to scatter light is often expressed by the haze ratio, which is defined as $H = T_{\text{diffuse}}/T_{\text{total}}$, where $T_{\text{total}} = T_{\text{diffuse}} + T_{\text{specular}}$. The measured value depends on the medium covering the TCO layer, but it is usually taken in air. The transmission data used are a weighted average over the spectral region of interest. It should be noted that TCO layers with different morphologies can have the same numeric value for the haze ratio. Therefore, the haze value has only limited significance as a figure of merit.

The use of textured TCO to achieve optically enhanced a-Si:H solar cells was first described by Deckman et al., 1983. The theoretical maximum enhancement factor for the optical path length is $4n^2$, where n is the refractive index of the medium (Yablonovitch and Cody, 1982). However, an enhancement factor of 40 - 50 has never been achieved in practice, and has experimentally been determined to be about 5 (Walker et al., 1987). In practice, the haze ratio increases with increasing thickness of the TCO. A low haze ratio of H = 6 - 12 % is appropriate for considerable absorption enhancement. This is due to the fact that the roughness of the TCO surface is to a large extent replicated at the interface of the n-layer of the p^+-i-n^+ structure and the back metal electrode, so that there is also effective scattering at this mirror-like interface.

The most frequently used preparation techniques for TCO layers are CVD, spray pyrolysis, (reactive) thermal evaporation, and (reactive) sputtering. It appears that for each type of metal oxide a different preparation technique dominates. A comprehensive review of TCO materials has been made by Hartnagel et al., 1995. Tin oxide, SnO_2, is usually made by Atmospheric Pressure CVD (APCVD) (Iida et al., 1983, Mizuhashi et al., 1988, Gordon et al., 1989), which produces a natively textured coating. Electron microscope graphs of this type of TCO are shown in Fig. 4.3 and Fig. 4.4.

The columnar crystals are nearly ideal with their base diameter of 200 nm and a pyramidal morphology at the surface exhibiting well-defined crystalline facets. This material shows excellent chemical stability in a reactive hydrogen plasma, although it does not withstand the strongly reducing ambient during HWCVD (Wallinga, 1998). The deposition temperature for this process is 500 - 600 °C, which basically limits the choice of superstrate materials to glass. Indium tin oxide, ITO, is usually made by sputtering or thermal evaporation (Hamberg and Granqvist, 1986). This yields highly conductive polycrystalline but comparatively smooth films consisting of small grains. The

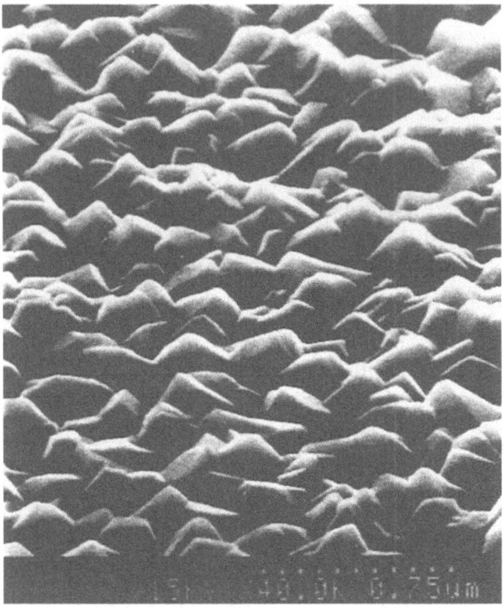

Figure 4.3 Scanning Electron Micrograph of Asahi type-U TCO (By courtesy of K. Adachi, Asahi Glass Company).

Table 4.1. Typical properties for various TCO front electrode coatings

Property	*Requirement*	*ITO*	SnO_2*:F*	*ZnO:Al*
transmission (%)	> 85	95	90	90
band gap (eV)	> 3.5	3.7	4.3	3.4
sheet resistance (Ω/□)	< 10*	3 -5	6 - 15	6 - 15
contact resistance to amorphous p^+-layer	low ohmic	low ohmic	low ohmic	forms barrier[b]
roughness	textured	negligible	excellent	negligible[c]
plasma durability	resistant	low[a]	good	excellent

* For single junction cells; can be relaxed for multijunction cells.
[a] A protective transparent coating is required to avoid severe reduction and diffusion of In. These layers need not be conductive, so apart from ZnO and SnO_2, materials as TiO_2 (Daey Ouwens et al., 1994) and SiO (De Nijs et al., 1991) were found suitable.
[b] The barrier can be eliminated by the use of microcrystalline doped layers.
[c] A textured surface can be formed after deposition by a suitable texture etch.

optimum deposition temperature is 200 - 250 °C. Zinc oxide (ZnO) is usually made by sputtering from ceramic targets (Schropp and Madan, 1989) or Metal

Figure 4.4. Cross-sectional Transmission Electron Micrograph of Asahi type-U TCO (By courtesy of K. Adachi, Asahi Glass Company).

Organic CVD (MOCVD) (Roth and Williams, 1981). These methods also result in quite smooth films with a high optical transmission and a conductivity that is slightly less than that of ITO. The deposition temperatures reported vary from room temperature to about 300 °C. The deposition rate using the sputtering technique can be enhanced by using dc sputtering rather than rf sputtering.

Sputtered ZnO films can be texture-etched in dilute HCl to produce a rough scattering surface. This technique has been demonstrated to be suitable for a-Si:H solar cells (Kluth et al., 1997).

The main properties of various front TCO electrode coatings are summarized in Table 4.1.

4.2.2 Semiconductor multilayer structure

In amorphous silicon p^+-i-n^+ superstrate type cells, the construction of the front 10 - 20 nm has a dominant influence on the overall performance. In 1981, a major improvement in the efficiency was obtained by incorporating a carbonated wide band gap p-layer, thus improving its window properties (Tawada et al., 1981). The properties of such p-layers have been discussed in Section 3.2.1. Arya et al., 1986 introduced the use of intentionally graded "buffer" p^+/i interfaces to enhance the performance in the blue region of the spectrum. The window properties of the p^+-layer have been improved further by using profiling schemes within the p^+-layer (Miyachi et al., 1992).

There have been a number of explanations for the efficiency enhancement due to the p^+/i buffer layer. The most frequently found explanations are: (i) the classical explanation of relaxation of the band gap discontinuity by reducing the bond distortion due to lattice strain at the interface, (ii) prevention of excessive boron diffusion from the p-layer, (iii) prevention of back diffusion of photogenerated electrons, (iv) spreading of the electric field into the i-layer near the p^+/i interface, and (v) prevention of recombination by spatially separating photogenerated electrons from their positive counterpart. In general, the improved quantum efficiencies in the blue region and the higher open circuit voltage and fill factor have usually been attributed to a reduced interface density of recombination centers near the junction (Arya et al., 1986, Yamanaka et al., 1987). However, the reduced recombination losses can also be interpreted as due to the prevention of access to interfacial defects rather than to actual defect minimization (Von Roedern, 1992).

The deposition of microcrystalline p-layers has been discussed in Section 3.2.2 and was for the first time successfully applied in a superstrate structure directly on SnO_2:F by Rath and Schropp, 1998.

The properties of the n-layer and the i/n interface are much less critical to the cell performance because the local generation rate is much smaller than at the p/i interface. The n-layer near the back contact is often made microcrystalline as this has an advantageous influence on the built-in voltage as well as on the contact resistance. Introducing a wide band gap or a lightly doped i/n buffer layer in this case has been reported to have beneficial effects (Tanaka et al, 1993).

4.2.3 Back electrode technology

In a superstrate pin solar cell, the roughness of the interface between the amorphous silicon and the metal back reflector contributes to the light trapping. Figure 4.5 shows the roughness of a 500 nm thick complete a-Si:H pin multilayer structure deposited on a textured TCO layer such as the one shown in Fig. 4.2.1. This roughness replicates the texture of the front electrode, though it is considerably smoothened. In Chapter 7, the effect of the rough interfaces at the front and back electrodes on the optical absorption is described in detail. In order to further enhance the light trapping within the a-Si:H solar cell, the reflectivity of the metal electrode is often increased by introducing a TCO layer (either ZnO or ITO) between the n-layer and the metal back contact. This idea was patented by Carlson and Williams, 1984. The large difference between the refractive index of the transparent conductor and the metal, combined with the

Figure 4.5. Scanning Electron Micrograph of a 500 nm thick complete a-Si:H pin multi-layer structure deposited on a textured TCO layer.

oblique incidence of a large fraction of the light at this interface, leads to total internal reflection for this fraction, and thus a high overall reflectivity. The gain in the short circuit current density J_{sc} can be as much as 1 mA/cm^2, which is mainly due to enhanced absorption in the long wavelength region (600 - 800 nm), as shown, for example, by Morris et al., 1990 and Beneking et al., 1994. The optimum thickness for the TCO layer is 70 nm which has been derived from optical modeling (Tao et al, 1992) and this value is confirmed by experiments (Landweer et al, 1994). In Section 8.1.3 these optical enhancement effects are further elaborated.

As ZnO is more transparent than ITO, and because the specific resisitivity requirement is substantially relaxed due to the small optimum thickness that is required, ZnO is the most commonly used reflection enhancement layer. However, since ZnO commonly has a large contact resistance with doped amorphous silicon (Kubon et al, 1996), the use of a microcrystalline layer is required. In an optimized cell using ZnO deposited by Metal-Organic CVD (MOCVD), Wenas et al., 1994 achieved an initial efficiency of 12.5 % in a single junction cell.

4.3 SUBSTRATE SOLAR CELLS

In substrate-type solar cells, the carrier on which the various thin film materials are deposited forms the back side of the cell. They are usually made on a stainless steel carrier that serves at the same time as the back contact. A highly flexible device can be made in the substrate configuration by employing a very thin metal carrier or a metal coated polymer foil. Since the polymer need not be transparent a temperature resistant type of polymer can be employed such as polyimid.

As an example, a cross section of a triple junction substrate solar cell made on stainless steel is shown in Fig. 4.6.

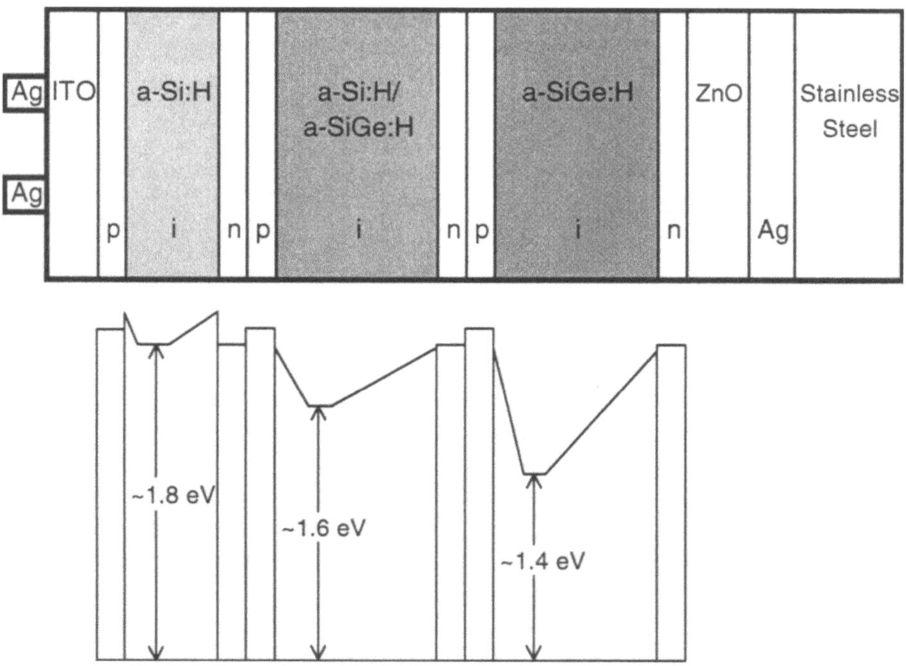

Figure 4.6. Cross-sectional diagram of a substrate triple junction solar cell and corresponding schematic band diagram.

4.3.1 Back electrode technology

In order to enhance the back reflecting properties of the back contact, the stainless steel carrier can be coated with a Ag/ZnO bilayer to achieve similar reflectance as in the superstrate structure. The ZnO may have an additional function in that it prevents interdiffusion of the Ag and the n-layer. Further, the evaporated Ag layer can be made textured to enhance the effective optical path-length. High temperatures can be used for the deposition of Ag and/or ZnO to obtain the desired texture, since no silicon layers are present at the stage of forming the back reflector. The use of a textured Ag/ZnO by Banerjee and Guha, 1991 resulted in a short circuit current density of as much as 17.5 mA/cm^2 in a single junction cell. This is about 6 mA/cm^2 higher than in a single junction cell on plain stainless steel. From photothermal deflection spectroscopy (PDS) it was concluded that a further gain must theoretically be achievable (Deng and Narasimhan, 1994). At Kaneka Corp., Japan, a thin film polycrystalline n-i-p cell has recently been developed with an aperture area efficiency of 10.1 % (Yamamoto et al., 1998). Due to the utilization of a highly textured back reflector, a plasma-deposited polycrystalline absorber layer, which is merely 2 μm thick, was shown to be capable of generating as much as 25.9 mA/cm^2 under short-circuit conditions.

4.3.2 Semiconductor multilayer structure

After the n-type layer, a buffer layer can be deposited similar to the i/n interface layer in p-i-n structures. If the n-layer is microcrystalline this buffer layer is necessary to bridge the band offset that is likely to exist between the microcrystalline n-layer and the amorphous i-layer, however, if the n-layer is amorphous a wide band gap transition layer might be advantageous as well. The i-layer can be made of device-quality amorphous, micro- or polycrystalline material.

An advantage of the n-i-p structure is that deposition temperature grading to adjust the band gap profile can be much more easily implemented as the band gap in amorphous hydrogenated semiconductors generally increases at decreasing deposition temperatures. A further advantage is that the most critical top electrode/p-type window layers are deposited last and need not be exposed to subsequent plasmas combined with elevated temperatures. In addition, microcrystalline p-layers are more successful in a substrate configuration as they are deposited on a silicon layer instead of on a metal oxide layer where the risk of chemical reduction is large.

In order to obtain good crystallinity of the p-layer, as well as an appropriate band-offset engineered transition, it is important to deposit a carefully optimized buffer layer before the microcrystalline p-layer.

4.3.3 Front electrode technology

In the n-i-p configuration the transparent top contact is deposited last. This imposes a restriction to the deposition temperature that can be used. Moreover, the p-layer has to form a tunnel contact with the degenerately n-type doped metal oxide layer, which imposes a high conductivity requirement on the metal oxide layer. The most suitable top contact materials are evaporated ITO or ZnO deposited by the MOCVD technique. The thickness of this layer is chosen such that it provides an antireflection layer. As this thickness is about 70-80 nm the sheet resistance of the best ITO is in practice still quite high (> 50 $\Omega/\Box$), and therefore a metal grid is necessary to reduce the series resistance and thus enhance the fill factor.

4.4 MULTIJUNCTION TECHNOLOGY

If two or more cells with different optical band gap are stacked on top of each other the conversion efficiency can be much improved compared to a single cell. This is due to the more efficient utilization of the energy per absorbed photon as well as to the improved collection of carriers. The latter advantage is due to the fact that each component cell has a thickness that is smaller than that of a typical single junction cell. In addition, these multijunction cells are less sensitive to light-induced degradation as each component cell can be made thin enough to avoid deterioration of the collection due to light-induced defect states. Therefore, stacked cells have stabilized efficiencies superior to single junction cells even if the additional cell has the same band gap as the first cell.

Optimized series-connected stacked multijunctions are not simply a repetition of the p^+-i-n^+ or n^+-i-p^+ structures on top of each other. Several issues require special attention, among others:

- The generation profile in the second (or third) junction from the top is much more uniform that that in the first junction, due to the filtering effect of the first junction. Mainly the blue light has been filtered out and the absorption coefficients for the remaining light are much lower leading to a more uniform photocarrier generation.

- The component cells deeper in the stack can be made with a-SiGe:H alloys that have a band gap lower than a-Si:H. As the transport properties of these alloys are not as good as those of a-Si:H, a counterintuitive band-gap engineering scheme yields the best results. A tapered band-gap grading scheme, first introduced by Guha et al., 1989, starting from the top with a decreasing band gap and comprising an increasing band gap over the remaining (major) part of the absorber layer provides the right balance between the absorption profile (for carrier generation) and internal field profile (for hole transport).

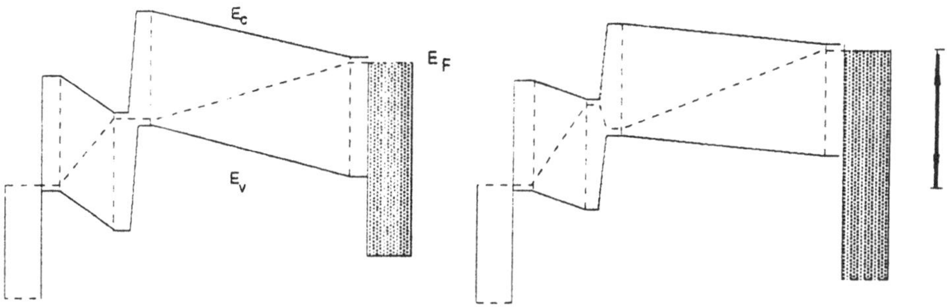

Figure 4.7. Multijunction cell under forward bias voltage, **left:** with a good tunnel-recombination junction **right:** showing the potential loss of built-in voltage due to an inadequate tunnel-recombination junction.

- Microcrystalline doped layers are playing an important role in two-terminal multijunction solar cells, where they provide a high built-in voltage and low optical losses. At the internal junctions of multijunction cells they have the additional function of providing an efficient recombination channel for oppositely charged carriers from both adjacent junctions. If the recombination rate at the multicell interface does not keep up with the supply or carriers to this region, space charge accumulation will be the result which adversely affects the field in the adjacent cell that has the highest current generation. In Fig. 4.7 the loss mechanism due to an inadequate tunnel-recombination junction (TRJ) is schematically depicted. Most tunnel-recombination junctions (TRJs) consist of one microcrystalline layer and one amorphous layer. In principle, an all-microcrystalline TRJ would offer the lowest optical losses. Much effort is put into maintaining sufficient carrier recombination at TRJs consisting of two oppositely doped microcrystalline layers to avoid charge accumulation at the tunnel-recombinaton junction (Rath et al., 1997, Yang et al., 1997).

- The current density in all component cells of the multijunction device has to be equal, since the output current is dictated by the lowest current generated in the series-connected stack. This requires a combined modeling and empirical approach for optimizing the efficiency under various conditions, i.e. including internal light scattering, cell-to-cell band gap variation, properties of the TRJs (also see Section 8.3.2), band gap profiling within each cell, load conditions, spectrum changes, and the changes due to light-induced effects in each cell.

Since the discovery that good wide band gap material could be produced without alloying with carbon, but merely by hydrogen dilution during deposition (Section 2.1.3.2), most triple junction cells nowadays have an a-Si:H i-layer in the top cell and a-SiGe:H i-layers in the middle and the bottom cell.

The highest stabilized efficiency to date has been achieved at United Solar (Yang et al., 1997) with a triple junction structure in the substrate configuration. If the active area efficiency is calculated by subtracting the area shaded by the grid from the total cell area, the stabilized active area efficiency was 13.0 %. Here, *stabilized efficiency* is defined as the efficiency after 1000 hours exposure at 1 sun intensity at 50 °C. The initial active area efficiency for this cell was 14.5 %. Further improvements in the stabilized efficiency can be expected, as a triple junction cell with an initial efficiency of 15.2 % was recently reported (Yang et al., 1998). The solar cell parameters pertaining to the confirmed stabilized total area efficiency of 12.1 % are listed in Table 4.2, along with the parameters of competitive small area cells. In order to achieve this result, special attention was paid to (i) the transmission of the ITO, (ii) the incorporation of both microcrystalline n- and p-layers at the TRJs, along with appropriate buffer layers before the microcrystalline n-layer and between the n- and p-layer, (iii) high hydrogen dilution during deposition of all three intrinsic

Table 4.2. Record stabilized laboratory cell aperture-area efficiencies.

Cell type	*Organization*	*Area* (cm^2)	V_{oc} *(V)*	J_{sc} $(mA.cm^{-2})$	*FF* -	*Efficiency* *(%)*
a-Si	APS[1]	1.02	0.854	16.52	0.615	8.54*
a-Si	Sanyo[2]	1	0.86	15.8	0.65	8.8*
a-Si	United Solar[3]	0.25	0.965	14.36	0.672	9.3†,*
a-Si/a-SiGe	Sanyo[2]	1	1.49	10.9	0.654	10.6*
a-Si/a-SiGe/a-SiGe	United Solar[4]	0.25	2.30	7.56	0.70	12.1*
a-Si/μc-Si[a]	Neuchâtel[5]	0.25	1.284	13.5	0.692	12.0**
poly-Si	Kaneka[6]	1.2	0.539	24.35	0.768	10.1***
μc-Si	Neuchâtel[7]	0.25	0.531	22.9	0.698	8.5***

Literature references: [1] Xi et al., 1994, [2] Hishikawa et al., 1994, [3] Guha et al., 1998, [4] Yang et al., 1997, [5] Meier et al., 1997, [6] Yamamoto et al., 1998, [7] Meier et al., 1998

* Stabilized efficiency, light-soaked indoors

† Active area, corrected for shadow loss

** Stabilized efficiency, measured outdoors

*** Not light-soaked, but not suffering light-induced degradation

[a] "Micromorph" concept

layers along with an increased Ge content in the bottom cell. Previous optimizations already addressed the band-gap profiling and the textured Ag/ZnO back contact. The values of the respective band gaps were not disclosed, but according to computer simulations by Rochelau et al., 1997 the Tauc optical band gaps were 1.79 eV, 1.55 eV and 1.39 eV, from top to bottom. In practice, however, due to band-gap profiling within the bulk of each i-layer and at the interfaces, none of the component cells has a single value for the band gap.

A promising multibandgap concept is the so called micromorph cell concept (Meier et al., 1998), in which amorphous ($E_g \approx 1.7$ eV) and microcrystalline ($E_g = 1.1$ eV) silicon are combined. Apart from the fact that this band-gap combination is almost optimal for the terrestrial solar spectrum, the advantages of this structure are that it circumvents the difficulties in achieving high quality a-SiGe:H with a Tauc optical band gap lower than 1.5 eV (Section 3.1.3) and that the microcrystalline component cell can be made stable. The efforts of the Neuchâtel group so far have led to a confirmed stabilized efficiency of 10.7 % (stabilized after 1000 hours of light soaking) (Keppner et al., 1996), and an unconfirmed stabilized efficiency of 12.0 % (outdoor measurement; Shah, 1998). The issues under investigation are the low deposition rate of μc-Si:H, and the scaling up of the deposition of this material over areas of industrial scale. A further concern is that the amorphous component cell still shows light-induced degradation, and therefore a slightly reduced band gap of this cell is desirable in order to allow for a thickness reduction.

4.5 FABRICATION OF LARGE AREA MODULES

The various methods for large area deposition of silicon thin films have been discussed in Section 2.1.1. This section describes the manufacturing of solar modules in which the silicon deposition process is integrated.

Stacked cells can be series- or parallel-connected. We will discuss both design schemes.

4.5.1 Series-connected stacked multijunctions

In order to reduce the series resistance losses that inevitably occur upon scaling up the area that a cell occupies, most commercial a-Si:H modules are constructed with monolithically interconnected cells. The series connection of the individual cells on the same substrate is achieved by including two or more laser patterning steps between the layer deposition steps. The laser cutting steps selectively remove narrow regions (50 - 150 μm) of a specific layer, ideally without affecting previously deposited layers. In some cases the laser beam is not used to remove material, but to thoroughly mix a number of layers in a well-defined area (laser *welding*), such that a local short circuit is created from one electrode to the opposite electrode establishing a series connection of two neighbouring cells. The area loss due to tolerances in the glass dimensions and registration inaccuracies amounts to 10-15 % in production. The output performance of modules can be greatly improved by accurate alignment of the subsequent laser scribing steps and by tighter control of the laser parameters in order to minimize the edge damage along the patterned zone. The output power of modules is also considerably improved when advancing from single junction to tandem or triple junction technology, even if the the cel efficiency is the same, merely by the reduction of interconnect area losses. This is because the current density generated in multijunction cells is considerably lower than in single junction cells of the same efficiency. This allows for the panel design to be constructed from wider subcells, so that the number of interconnect zones from one outer edge of the panel to the other is reduced and thus the fraction of inactive area is reduced.

Usually, Q-switched Nd:YAG lasers are used for laser patterning of the respective films. For scribing the front electrode TCO, the wavelength of 1.06 μm is very appropriate. For removing the silicon single junction or tandem p^{+}-i-n^{+} structure, or the metal back electrode, or both simultaneously, the frequency doubled mode ($\lambda = 0.53\ \mu$m) is used. Laser patterning can be carried out at a linear speed of 20 - 50 cm/s. Assuming a 1 m^2 panel size, with subcell divisions of 1 cm width, a single scribing step would take more than 3 minutes and could thus impose a throughput limitation in production. The solution to this is the use of either multiple laser stations or multiple beamsplitting of very powerful lasers.

Comparing typical small area laboratory cell efficiencies and equivalent large-area monolithically interconnected prototype panel efficiencies, usually a substantial gap in performance can be noted. This performance gap is due to the following issues:

- Interconnect area losses (as discussed above).
- Interconnect contact series resistance losses. Incomplete removal or redeposited debris can cause locally increased contact resistance between front and back contact layers at the interconnects.
- Layer nonuniformity. This can occur in any layer. Any local reduction of photovoltaic parameters (V_{oc}, J_{sc}, or FF) in the string of subcells will affect the entire panel. The performance of the p^{+}-i-n^{+} structure is most severely affected by variation in the p^{+}-layer thickness and/or properties. The built-in voltage (and thus V_{oc} and FF) is most sensitive to the p^{+}-layer nonuniformity. Also J_{sc} can be reduced if the p^{+}-layer becomes too thick. Specifically in multijunction cells, thickness non-uniformity in the intrinsic

layers can adversely affect the current matching properties and thus reduce the overall current output. Moreover, also the tunnel-recombination junction must have a constant thickness over the whole area. Finally, both in single junction and multijunction solar cells, nonuniformity in the TCO layer can locally alter the transmission or the light trapping capability.

- Difficulty in implementing band-gap and dopant engineering profiles over a large area. As an example, the buffer layer between the p^+-layer and the i-layer is compositionally profiled over a thickness of typically 2 - 5 nm. This profile greatly depends on the gas composition locally in the plasma. This composition may vary due to the flow geometry and the local depletion of the various constituents of the gas mixture. The formation of an adequate buffer layer thus requires that the gas composition is homogeneous over the whole area at all times during the deposition.

- Local shunting. Devices with a larger area have a larger probability to have a local shunting path incorporated in the device. These shunts may originate from (i) irregularities in the morphology of the TCO, (ii) particles due to the TCO deposition process, (iii) dust from the ambient, (iv) dust or flakes accumulated in the silicon deposition system, (v) edge damage due to laser patterning steps, (vi) dust incorporated due to gas phase nucleation in the CVD process.

Considering the above list of issues causing potential losses in scaling up, improving the quality of the laser patterning is only one of the necessary approaches to decrease the performance gap between typical small area cells and large area panels. Most issues can be overcome to a large extent as demon-

Table 4.3. Record stabilized total-area prototype module efficiencies.

Cell type	*Organization*	*Area* (cm^2)	V_{oc} (*V*)	I_{sc} (*mA*)	*FF* -	*Efficiency* %
a-Si/a-SiGe	Solarex[1]	842	36.17	347.5	0.609	9.09[a,*]
a-Si/a-SiGe	Sanyo[2]	1200	42.1	419	0.644	9.5 [b,*]
a-Si/a-SiGe/a-SiGe	United Solar[3]	903	2.318	6470	0.612	10.17[a,*]

Literature references: [1]Arya et al., 1994, [2]Terakawa et al., 1997, [3]Yang et al., 1994
* Stabilized efficiency, light-soaked indoors
[a]Verified at NREL (National Renewable Energy Laboratory, USA)
[b]Verified at JQA (Japan Quality Assurance Organization)

strated by Sanyo Electric Co. Comparing, for example, cells of the same type (stabilized a-Si/a-SiGe tandem cells, 10.6 % over 1 cm^2 and 9.5 % over 1200 cm^2), the loss can presently be limited to only 10 %.

In typical *substrate* modules on conducting carriers, by definition, monolithic interconnect techniques cannot be implemented. The way to reduce lateral series resistances in transparent contact electrodes is to use a metal grid. In scaling up, the same issues as above play a role, however the first two issues are then replaced by the issue of shadowing losses due to the grid. Further, additional active area losses may arise from the fact that individual subcell have to be placed side-by-side in a module with external connections in order to include them in a string. It is also of interest to note that, due to the higher operating voltage of triple junction cells than that of single or double junction cells, the advantage to utilizing multijunction cells is much larger for conductive substrate modules than for monolithically integrated superstrate modules.

Table 4.3 shows some of the highest stabilized efficiencies obtained for prototype modules.

4.5.2 Parallel-connected stacked multijunctions

In 1994, Siemens AG in Germany (Kusian et al., 1994) first proposed the parallel-connected stacked multijunction structure. Such a structure has the primary advantage that losses due to current mismatch can be avoided by incorporating a third terminal for current extraction, similar to a three-terminal tandem device. Therefore, non-uniformity in the intrinsic absorber layer thickness over larger areas is not as detrimental as in series-connected multijunction cells. At the same time the component junctions can be kept thin, which reduces the light-induced degradation of the structure due to Staebler-Wronski type recombination centers. Since the lateral conductance of thin doped amorphous or microcrystalline layer is low, the internal terminal between the two junctions of a stacked structure was made of indium tin oxide.

Whereas series-connected stacked cells suffer from limitation due to possible photocurrent mismatch, the drawback of the alternative parallel-connected structure is that the multitude of junctions should be voltage-matched. Since the operating voltage of a cell roughly scales with the optical band gap of the absorber, such a parallel-connected structure could never fully benefit from the advantages of enhanced utilization of the solar photon spectrum by incorporating different band-gap components. Further, since the various layers in such a structure can only be made by a large number of deposition steps, interrupted by numerous laser patterning steps, the cost-effectiveness of the method is also questionable.

A similar parallel-connected multijunction structure has also been proposed by UNSW in 1994 (Green et al., 1994). Instead of amorphous silicon films, alternatingly p- and n-doped polycrystalline thin films are interconnected by the buried contact technology. A multitude of junctions made of low-temperature deposited thin-film silicon materials can be stacked on a cheap substrate up to a total thickness of several tenths of micrometers in order to ensure sufficient light absorption, even when the electronic quality is not high enough to collect photogenerated carriers over such a thick layer. The ability to contact each component junction individually in parallel, allows the use of inexpensive deposited thin-film silicon materials. Several deposition methods, including the HWCVD and VHFCVD techniques discussed in this book, can be used for this concept.

In general however, if light trapping techniques such as surface texture and advanced back reflectors (Yamamoto et al., 1998) lead to highly efficient stable thin film single junction cells, one could question whether the stacked junction approach is still technologically and/or economically attractive.

4.5.3 Encapsulation

Module encapsulation is of critical importance since even the smallest traces of water penetration can cause severe corrosion or delamination of contact layers.

Superstrate modules built on glas superstrates have the best encapsulant imaginable on the window side of the panel. Water vapor or even ions originating from salt do not pass through a few mm thick glass pane at normal operating temperatures and the glass protects from environmental effects, such as hail. A second piece of glass is also the most effective way to encapsulate the rear side. It can be attached to the panel using resin or varnishes, such as polyurethane, or acrylic paint. There is no requirement that it should be transparent, neither is there any problem with eventual discoloration under the influence of light. Therefore, conventional EVA (ethyl-vinyl acetate copolymer) can be used as well. However, the weak spot is the panel edges where the two glas panes need to be joined hermetically in order to protect the interior of the module. Sealing of the circumference is a major issue which requires appropriate attention. The frame needs to be carefully designed and the framing material thoughtfully selected for optimum adhesion and compatibility of expansion coefficients. The seal usually improves by introducing a glass edge grinding step in the production sequence.

Substrate modules on conducting carriers as well as on polyimid foils have the drawback that they need to be encapsulated at both faces. For outdoor applications, the use of quite expensive materials is necessary, such as Tefzel (combined with EVA) and Tedlar (Dupont). Edge-sealing is often obtained by keeping the active materials about 2 cm away from all four outer edges. In this design, the total-area efficiency of the module is further reduced due to the large fraction of inactive area.

The current IEC-Standard 1646 test for commercial applications includes a few thorough tests as listed in Table 4.4.

Table 4.4. Current module testing procedure.

Test	*Requirement*
Thermal cycling	50 to 200 cycles from -40°C to +85°C
Humidity freeze	10 cycles from +85°C at 85 % relative humidity to -40°C
Damp heat	1000 hours at +85°C at 85 % relative humidity

4.6 PRODUCTION AND MARKETS FOR AMORPHOUS SILICON PHOTOVOLTAICS

As an example of a typical automated semi-continuous production line for the manufacturing of amorphous silicon superstrate modules on glass we here list the process flow of the 1 MWp/year line that Solarex started up in 1990. The production sequence used in more recently commissioned production lines at industries is not available for publication, but it should be expected that the new production lines for glass-superstrate panels do not differ basically from the one presented here.

As an example of a typical automated roll-to-roll production line for the manufacturing of amorphous silicon substrate modules on flexible foil such as polyimid we here discuss the 1 MWp/yr prototype line that Fuji Electric presented in 1995 (Fujikake et al., 1995). The process flow is listed in table 4.6. The line actually consists of a number of separate stations where the entire roll undergoes a partial process. Typical for the manufacturing process is the so called Stepping Roll (SR) film deposition and the Series-Connection through Apertures formed on Film (SCAF). The stepping roll apparatus has been designed to overcome interdiffusion of dopant gases and thus consists of a common chamber with separate deposition reactors in it.

Doped and undoped amorphous silicon layers are deposited simultaneously in the separated deposition zones. The deposition compartments are isolated from each other by pressing the O-ring seals of the compartments against a stationary foil.

Table 4.5. Description of a semi-continuous production line for a-Si:H modules.

Step	*Description*	*Purpose*
1	Glass edge abrasive treatment	Protect glass edges during subsequent processing
2	Washing	Remove debris
3	Deposition of 50 nm SiO_2 using APCVD	Achieve uniform haze of SnO_2 (and reduce sodium diffusion from glass)
4	Deposition of 600 nm SnO_2:F in the same belt furnace	Produce front electrode with proper texture
5	Test station	Check TCO resistance
6	Pasting & curing of Ag frit	Formation of highly conductive bus bars
7	Laser patterning	Isolate front contacts of subcells
8	Wet cleaning	Remove debris from laser patterning
9	Loading of pieces in PECVD multichamber system	Processing of multiple pieces
10[a]	Deposition of the p-layer	Formation of the doped contact layer
11[a]	Deposition of the i and the n-layer	Formation of the rest of the cell
12	Al deposition by sputtering	Formation of the back contact
13	Laser scribing through glass	Patterning of the back contact (silicon is also removed)
14	Laser scribing with higher power through glass	Isolation of the edges (all layers removed)
15	Flipping of the panel	Perform the following steps from the back side
16	Dry brushing	Removal of any residue
17	Laser welding of the Si/metal region between the TCO scribe and the metal scribe	Formation of the interconnect region
18	Electrical curing	Repair of shunting paths by burning out wherever possible
19	Performance test	Checking of the interconnected module
20	Spray-coating	Application of the encapsulating paint

[a]Steps 10 and 11 may be repeated for production of multijunction cells.

After completion of a deposition step, the compartments are opened and the foil is conveyed such that each previously coated region is positioned at the next compartment in the sequence. The "SCAF" method for making monolithic interconnects makes use of the fact that small holes can easily be made in a plastic foil. One set of holes is for current collection (to avoid the series resistance of the front ITO without the need of making a grid electrode at the front); another set of holes at the sides, where no front ITO is deposited, is for series connection of the subcells. By producing the layers in the correct order, current collection contacts and series connection contacts are automatically formed. Laser scribing is used to pattern the coatings on the front and back side such that subcells are defined.

Table 4.6. Description of a roll-to-roll production line for a-Si:H modules.

Step	*Description*	*Purpose*
1	Mechanical punching of holes in polyimid foil	Provide series-connection and current-collection holes
2	Cleaning	Remove debris
3	Sputter deposition of metal coating on "back" side	Formation of a continuous electrode for the collection of current, to avoid the sheet resistance of the ITO front electrode
4	Sputter deposition of metal coating on "front" side	Formation of the back electrode and prevention of outgasing from the foil
5	Deposition of the tandem cell in stepping mode fashion	Formation of the silicon solar cell structure with low levels of cross contamination
6	Deposition of the ITO front electrode	Provide the front electrode window coating
7	Sputter deposition of metal coating on "back" side	Formation of contacts between transparent ITO front electrode and back-side electrode
8	Laser patterning front and back side	Separation into subcells
9	Apply EVA and protection cover film to front and back side	Encapsulation; the same step is used for metal tape application
10	Roll-to-roll vacuum lamination	Completion of lamination step
11	Performance test	Checking of the interconnected module

Due to the fact that interconnects are formed by deposition into the holes, the difficulty of selective laser scribing of individual layers on a substrate foil as well as the requirement of precise control of the laser beam is avoided.

The total market for PV modules in 1997 was 127 MW (Maycock, 1998). The growth in the market was roughly 40 % with respect to 1996, and before that the annual growth rate used to be 15 %. The market share of amorphous silicon products is roughly 15-20 %. Cadmium telluride cells supply only 1-2 %, chalcopyrite cells (such as Cu(In,Ga)Se_2 cells) are not on the market as yet, but several production plans exist in Australia, U.S.A., Germany, and Japan. Dye-sensitized solar cells are currently introduced on a very small scale in solar watches.

Table 4.7. Production facilities for a-Si:H and related solar cells and modules.

Country	*Company name*	*Production*	*Typical products*
USA	ITFT	0.1 MW[b]	Architectural and recreational specialty products
	Solarex Corp.	10 MW[a]	PV systems
	United Solar	5 MW[a]	PV systems, roof integrated systems
Japan	Canon	10 MW[a]	
	Fuji Electric	1 MW[a]	Demonstration prototypes
	Kaneka Corp.		
	Kyocera Corp.		
	Sanyo Electric Co.	6.5 MW[b]	Consumer products and architectural applications
	Sharp Corp.	1 MW[a]	Modules
	Taiyo Yuden Co.		
India	Bharat Heavy Electrical Ltd.	0.6 MW[b]	PV systems
Germany	Phototronics Solartechnik GmbH	1 MW[a]	Architectural glass
China	Harbin	1 MW[a]	PV systems
UK	Intersolar	1 MW[a]	PV systems
Croatia	Koncar Solar Cells	0.6 MW[b]	Lighting and pumping systems
France	FEE France S.A.	0.6 MW[b]	Lighting systems
Russia	Sovlux/Quant	2 MW[b]	a-Si/a-SiGe, PV systems

[a]Production capacity, [b]Estimated production

In a study conducted for the European Commission (MUSICFM; Woodcock et al., 1997) one possible scenario for the near future was presented, in which a 22.5 % per annum growth rate (7.5 % faster than a "business as usual" scenario) would lead to the dominance of the market by thin film PV products. If either chalcopyrite cells or cadmium telluride cells were to take a large share of the potential 500 MWp thin-film market by 2005, the limited availability of indium or tellurium would make recycling technologies advisable. There are no limitations in the world's resources with thin film silicon PV technology.

In Table 4.7 we list the companies that are currently active in production or pilot production of cells and modules incorporating amorphous and/or microcrystalline silicon.

References

Arya, R.R., A. Catalano, and R.S. Oswald, *Amorphous silicon p-i-n solar cells with graded interface*, Appl. Phys. Lett. **49** (1986) 1089-1091.

Arya, R.R., R.S. Oswald, Y.M. Li, N. Maley, K. Jansen, L. Yang, L.F. Chen, F. Willig, M.S. Bennet, J. Morris, and D.E. Carlson, *Progress in amorphous silicon based multijunction modules*, 1st World Conference on Photovoltaic Energy Conversion, (Proc. 24th IEEE PV Specialists Conference, Waikoloa, HI, USA, December 1994) 394-400.

Ashida, Y., *Single-junction a-Si solar cells with over 13 % efficiency*, Techn. Digest of the International PVSEC-7, Nagoya, Japan, 1993, 33-36.

Banerjee, A., and S. Guha, *Study of back reflectors for amorphous silicon alloy solar cell application*, J. Appl. Phys. **69** (1991) 1030-1035.

Beneking, C., B. Rech, Th. Eickhoff, Y.G. Michael, N. Schultz, and H. Wagner, *Preparation and light stability of a-Si/a-Si stacked solar cells*, 12th International E.C. Photovoltaic Solar Energy Conference 1994, Eds. R. Hill, W. Palz, and P. Helm (H.S. Stephens and Associates, 1994) 683-686.

Carlson, D.E., and C.R. Wronski, *Amorphous silicon solar cell*, Appl. Phys. Lett. **28** (1976) 671-673.

Carlson, D.E. and B.F. Williams, *Photodetector having enhanced back reflection*, U.S. Patent No. 4,442,310; April 10, 1984.

Daey Ouwens, J., R.E.I. Schropp, J. Wallinga, W.F. van der Weg, M. Ritala, M. Leskelä, and J. Hyvärinen, *Titanium dioxide as superior transparant conducting oxide for an improved conversion efficiency of a-Si solar cells*, 12th International E.C. Photovoltaic Solar Energy Conference 1994, Eds. R. Hill, W. Palz, and P. Helm (H.S. Stephens and Assoc., 1994) 1296-1299.

Deckman, H.W., C.R. Wronski, H. Witzke, and E. Yablonovitch, *Optically enhanced amorphous silicon solar cells*, Appl. Phys. Lett. **42** (1983) 968-970.

Deng, X., and K.L. Narasimhan, *New evaluation technique for thin-film solar cell back-reflector using photothermal deflection spectroscopy*, 1st World Conference on Photovoltaic Energy Conversion, (Proc. 24th IEEE PV Specialists Conference, Waikoloa, HI, USA, December 1994) 555-558.

De Nijs, J.M.M., C. Carvalho, M. Santos, and R. Martins, *A thin SiO layer as a remedy for the indium reduction at the* $In_2O_3/\mu c$*-Si:C:H interface*, Appl. Surf. Sci. **52** (1991) 339-342.

Fujikake, S., K. Tabuchi, T. Yoshida, Y. Ichikawa and H. Sakai, *Flexible a-Si solar cells with plastic film substrate*, in: Amorphous Silicon Technology - 1995, edited by M. Hack, E.A. Schiff, A. Madan, M. Powell, and A. Matsuda, Materials Research Society Symp. Proc. **377** (1995) 609-619.

Gordon, R., J. Proscia, F.B. Ellis, and A.E. Delahoy, *Textured tin oxide films produced by atmospheric pressure chemical vapor deposition from tetramethyltin and their usefulness ion producing light trapping in thin film amorphous silicon solar cells*, Solar Energy Mater. **18** (1989) 263-281.

Green, M.A., A. Wang, J. Zhao, G.F. Zheng, W. Zhang, Z. Shi, C.B. Honsberg, and S.R. Wenham, *23.5 % efficiency and other recent improvements in silicon solar cell and module performance*, 12th International E.C. Photovoltaic Solar Energy Conference 1994, Eds. R. Hill, W. Palz, and P. Helm (H.S. Stephens and Assoc., 1994) 776-779.

Guha, S., J. Yang, A. Pawlikiewicz, T. Glatfelter, R. Ross, and S.R. Ovshinsky, *Band-gap profiling for improving the efficiency of amorphous silicon alloy solar cells*, Appl. Phys. Lett. **54** (1989) 2330-2332.

Guha, S., J. Yang, A. Banerjee, and S. Sugiyama, *Material issues in the commercialization of amorphous silicon ally thin-film photovoltaic technology*, in: Amorphous and Microcrystalline Silicon Technology - 1998, edited by R. Schropp, H. Branz, S. Wagner, M. Hack, and I. Shimizu, Materials Research Society Symp. Proc. **507** (1998) in print.

Hack, M., and M. Shur, *Physics of amorphous silicon alloy p-i-n solar cells*, J. Appl. Phys. **58** (1985) 997-1020.

Hamberg, I., and C.G. Granqvist, *Evaporated Sn-doped In_2O_3 films: basic optical properties and applications to energy-efficient windows*, J. Appl. Phys. **60** (1986) R123-R159.

Hartnagel, H., A. Dawar, A. Jain, and C. Jagadish, *Semiconducting Transparent Thin Films*, (Institute of Physics Publishing, 1995).

Ichikawa, Y., T. Ihara, S. Saito, H. Ota, S. Fujikake, and H. Sakai, *Production technology for large area amorphous silicon solar cells with high efficiency*, 11th E.C. Photovoltaic Solar Energy Conference 1992, Eds. L. Guimarães, W. Palz, C. de Reyff, H. Kiess, and P. Helm (Harwood Academic Publishers, 1992) 203-206.

Iida, H., N. Shiba, T. Mishuku, H. Karasawa, A. Ito, M. Yamanaka, and Y. Hayashi, *Efficiency of the a-Si:H solar cell and grain size of SnO_2 transparent conductive film*, IEEE Electron Dev. Lett. **EDL-4** (1983) 157-159.

Keppner, H., P. Torres, J. Meier, R. Platz, D. Fischer, U. Kroll, S. Dubail, J.A. Anna Selvan, N. Pellaton Vaucher, Y. Ziegler, R. Tscharner, Ch. Hof, N. Beck, M. Goetz, P. Pernet, M. Goerlitzer, N. Wyrsch, J. Veuille, J. Cuperus, A. Shah, J. Pohl, *The "Micromorph" cell: a new way to high-efficiency low-temperature crystalline silicon thin-film cell manufacturing?*, in: Advances in Microcrystalline and Nanocrystalline Semiconductors - 1996, edited by R.W. Collins, P.M. Faucher, I. Shimizu, J.C. Vial, T. Shimada, and A.P. Alivisatos, Materials Research Society Symp. Proc. **452** (1996) 865-876.

Kluth, O., A. Löffl, S. Wieder, C. Beneking, W. Appenzeller, L. Houben, B. Rech, H. Wagner, S. Hoffmann, R. Waser, J.A. Anna Selvan, and H. Keppner, *Texture etched Al-doped ZnO: a new material for enhanced light trapping in thin film solar cells* Proc. of the 26th IEEE Photovoltaic Specialists Conference, 1997, Anaheim, CA, USA, 715-718.

Kubon, M., E. Böhmer, F. Siebke, B. Rech, C. Beneking, and H. Wagner, *Solution of the ZnO/p contact problem in a-Si:H solar cells*, Solar Energy Materials and Solar Cells **41/42** (1996) 485-492.

Kusian, W., J. Furlan, G. Conte, F. Smole, M. Topič, and P. Popovič, *The pin/TCO/-nip a-Si:H solar module*, 12th International E.C. Photovoltaic Solar Energy Conference 1994, Eds. R. Hill, W. Palz, and P. Helm (H.S. Stephens and Assoc., 1994) 1249-1252.

Landweer, G.E.N., B.S. Girwar, C.H.M. van der Werf, J.W. Metselaar, and R.E.I. Schropp, *Enhanced efficiency and stability of amorphous silicon tandem solar cells by applying a highly reflective back contact*, 12th International E.C. Photovoltaic Solar Energy Conference 1992, Eds. R. Hill, W. Palz, and P. Helm (H.S. Stephens and Associates, 1994) 1300-1303.

Maycock, P., PV News, February (1998).

Meier, J., S. Dubail, J. Cuperus, U. Kroll, R. Platz, P. Torres, J.A. Anna Selvan, P. Pernet, N. Beck, N. Pellaton Vaucher, Ch. Hof, D. Fischer, H. Keppner, and A. Shah, *Recent progress in micromorph cells*, J. Non-Cryst. Solids **227-230** (1998) 1250-1256.

Meier, J., H. Keppner, S. Dubail, U. Kroll, P. Torres, P. Pernet, Y. Ziegler, J.A. Anna Selvan, J. Cuperus, D. Fischer, and A. Shah, *Microcrystalline single-junction and micromorph tandem thin film silicon solar cells*, in: Amorphous and Microcrystalline Silicon Technology - 1998, edited by R. Schropp, H. Branz, S. Wagner, M. Hack, and I. Shimizu, Materials Research Society Symp. Proc. **507** (1998) in print.

Miyachi, K., N. Ishiguro, T. Miyashita, N. Yanagawa, H. Tanaka, M. Koyama, Y. Ashida, and N. Fukuda, *Fabrication of single pin type solar cells with a high conversion efficiency of 13.0 %*, 11th E.C. Photovoltaic Solar Energy Conference 1992, Eds. L. Guimarães, W. Palz, C. de Reyff, H. Kiess, and P. Helm (Harwood Academic Publishers, 1992) 88-91.

Mizuhashi, M., Y. Gotoh, and K. Adachi, *Texture morphology of SnO_2:F films and cell reflectance*, Jpn. J. Appl. Phys. **27** (1988) 2053-2061.

Morris, J., R.R. Arya, J.G. O'Dowd, and S. Wiedemann, *Absorption Enhancement in Hydrogenated Amorphous Silicon (a-Si:H) based solar cells*, J. Appl. Phys. **67** (1990) 1079-1087.

Hishikawa, Y., K. Ninomiya, E. Maruyama, S. Kuroda, A. Terakawa, K. Sayama, H. Tarui, M. Sasaki, S. Tsuda, and S. Nakano, *Approaches for stable multi-junction a-Si solar cells*, 1st World Conference on Photovoltaic Energy Conversion, (Proc. 24th IEEE PV Specialists Conference, Waikoloa, HI, USA, December 1994) 386-393.

Rath, J.K., F.A. Rubinelli, and R.E.I. Schropp, *Microcrystalline n- and p-layers at the tunnel junction of a-Si:H/a-Si:H tandem cells*, J. Non-Cryst. Solids **227-230** (1998) 1282-1286.

Rath, J.K., and R.E.I. Schropp, *Incorporation of p-type microcrystalline silicon films in amorphous silicon based solar cells in a superstrate structure*, Solar Energy Materials and Solar Cells **53** (1998) 189-203.

Rochelau, R.E., M. Tun, and S.S. Hegedus, *Analysis and optimization of high efficiency multijunction a-Si:H solar cells*, Proc. 26th IEEE PV Specialists Conference, Anaheim, CA, USA, 1997) 703-706.

Roth, A.P., and D.F. Williams, *Semiconducting ZnO fims prepared by Metal Organic CVD from diethyl zinc*, J. Electrochem. Soc. **128** (1981) 2684-2686.

Schropp, R.E.I., and A. Madan, *Properties of conductive zinc oxide films prepared by rf magnetron sputtering for transparent electrode applications* J. Appl. Phys. **66** (1989) 2027-2031.

Shah, A., *New and enhanced silicon solar cells*, First JOULE III PV Contractor's meeting, May 5-7, 1998, CCAB-Brussels.

Street, R.A., *Hydrogenated amorphous silicon*, (Cambridge University Press, Cambridge, U.K., 1991).

Tanaka, H., N. Ishiguro, T. Miyashita, N. Yanagawa, M. Sadamoto, M. Koyama, Y. Ashida, and N. Fukuda, *Fabricating high performance a-Si solar cells by alternately repeating deposition and hydrogen plasma treatment method*, Techn. Digest of the International PVSEC-7, Nagoya, Japan, 1993, 269-271.

Tao, G., B.S. Girwar, G. Landweer, M. Zeman, and J.W. Metselaar, *Highly reflective TCO/Al back contact for a-Si:H solar cells*, 11th E.C. Photovoltaic Solar Energy Conference 1992, Eds. L. Guimarães, W. Palz, C. de Reyff, H. Kiess, and P. Helm (Harwood Academic Publishers, 1992) 605-608.

Tawada, Y., H. Okamoto, and Y. Hamakawa, *a-SiC:H/a-Si:H heterojunction solar cell having more than 7.1 % efficiency*, Appl. Phys. Lett. **39** (1981) 237-239.

Terakawa, A., M. Shima, T. Kinoshita, M. Isomura, M. Tanaka, S. Kiyama, S. Tsuda, and H. Matsunami, *The effect of the optical gap and compositions of a-SiGe:H solar cells on the time decay of light-induced degradation*, Proc. of the 14th E.C. Photovoltaic Solar Energy Conference, 30 June - 4 July 1997, Barcelona, Spain, Eds. H.A. Ossenbrink, P. Helm, and H. Ehmann (H.S. Stephens and Associates, 1997) 2359-2362.

Von der Linden, M.B., R.E.I. Schropp, J. Stammeijer and W.F. van der Weg, *Evaluation of the DICE method as a tool for amorphous silicon solar cell optimization*, in: Amorphous Silicon Technology - 1992, edited by M.J. Thompson, Y. Hamakawa, P.G. LeComber, A. Madan, and E. Schiff, Materials Research Society Symp. Proc. **258** (1992) 935-940.

Von Roedern, B., *Higher cell efficiencies through defect engineering of solar cell junctions*, 11th E.C. Photovoltaic Solar Energy Conference 1992, Eds. L. Guimarães, W. Palz, C. de Reyff, H. Kiess, and P. Helm (Harwood Academic Publishers, 1992) 295-298.

Walker, C., R.E. Hollingsworth, and A. Madan, *Determination of the efficiency enhancement due to scattering from rough TCO contact for a-Si:H p-i-n solar cells*, in: Amorphous Silicon Semiconductors - Pure and Hydrogenated, edited by A. Madan, M. Thompson, D. Adler, and Y. Hamakawa, Materials Research Society Symp. Proc. **95** (1987) 527-532.

Wallinga, J., W.M. Arnold Bik, A.M. Vredenberg, R.E.I. Schropp, and W.F. van der Weg, *Reduction of tin oxide by hydrogen radicals*, to be published in J. Phys. Chem. (1998).

Wenas, W.W., K. Dairiki, A. Yamada, M. Konagai, K. Takahashi, J.H. Jang, and K.S. Lim, *High efficiency a-Si solar cells with ZnO films*, 1st World Conference on Photovoltaic Energy Conversion, (Proc. 24th IEEE PV Specialists Conference, Waikoloa, HI, USA, December 1994) 413-416.

Woodcock, J.M. H. Schade, H. Maurus, B. Dimmler, J. Springer, and A. Ricaud, *A study of the upscaling of thin film solar cell manufacture towards 500 MWp per annum*, Proc. of the 14th E.C. Photovoltaic Solar Energy Conference, 30 June -

4 July 1997, Barcelona, Spain, Eds. H.A. Ossenbrink, P. Helm, and H. Ehmann (H.S. Stephens and Associates, 1997) 857-860.

Xi, J., D. Shugar, and H. Volltrauer, *Large area module performance and identification and control of p-i interface-correlated device degradation and further improvement in stabilized efficiencies of single-junction a-Si solar cells*, 1st World Conference on Photovoltaic Energy Conversion, (Proc. 24th IEEE PV Specialists Conference, Waikoloa, HI, USA, December 1994) 401-404.

Yablanovitch, E., and G.D. Cody, *Intensity enhancement in textured optical sheets for solar cells*, IEEE Trans. Elec. Dev. **ED-29** (1982) 300-305.

Yamamoto, K. *Thin film poly-Si solar cell on glass substrate fabricated at low temperature*, in: Amorphous and Microcrystalline Silicon Technology - 1998, edited by R. Schropp, H. Branz, S. Wagner, M. Hack, and I. Shimizu, Materials Research Society Symp. Proc. **507** (1998) in print.

Yamanaka, S., S. Kawamura, M. Konagai and K. Takahashi, Technical Digest of the International PVSEC-3, Tokyo, Japan, 1987, 709-712.

Yang, J., A. Banerjee, T. Glatfelter, K. Hoffman, X. Xu, and S. Guha, *Progress in triple-junction amorphous silicon-based alloy solar cells and modules using hydrogen dilution*, 1st World Conference on Photovoltaic Energy Conversion, (Proc. 24th IEEE PV Specialists Conference, Waikoloa, HI, USA, December 1994) 380-385.

Yang, J., A. Banerjee, and S. Guha, *Triple-junction amorphous silicon alloy solar cell with 14.6 % initial and 13.0 % stable conversion efficiencies*, Appl. Phys. Lett. **70** (1997) 2975-2977.

Yang, J., A. Banerjee, S. Sugiyama, and S. Guha, *Correlation of component cells with high efficiency amorphous silicon alloy triple-junction solar cells and modules*, presented at the 2nd World Conference and Exhibition on Photovoltaic Energy Conversion, to be published.

5 METASTABILITY

I honestly can't remember the moment when I realized that there was something funny going on, but Chris remembers I accused him (jokingly I'm sure) that he couldn't measure these samples right.

—David L. Staebler, April 3, 1997

5.1 LIGHT-INDUCED CHANGES IN FILMS

The light-induced metastability of hydrogenated amorphous silicon was discovered in 1977 (Staebler and Wronski, 1977). The thin films deposited with the equipment available in those days were slightly n-type, and the light-induced change, later called the Staebler-Wronski effect (SWE), manifested itself in these materials as a shift of the Fermi level towards mid-gap accompanied by a reduction of the dark conductivity and the photoconductivity. In 1980, it was established that the SWE is a bulk effect rather than a surface band bending effect (Staebler and Wronski, 1980). Since then, all experimental data have shown evidence that exposure of hydrogenated amorphous silicon to light increases the density of neutral silicon dangling bonds. The excess defects, which are metastable as they can be removed in 1 - 3 hours by thermal annealing above 150 °C, have a roughly one order of magnitude higher concentration than the as-deposited dangling bonds that are initially present in device-quality material, and thus significantly reduce the lifetime of free carriers. The vast body of experiments carried out during the last 20 years by numerous laboratories suggests that the creation of metastable dangling bonds is the result of recombination events between carriers created by light absorption or by injection in the dark. Although it takes 10 - 100 million recombination events to create only a single metastable dangling bond, the concentration of $\approx 10^{17}$ cm^{-3} defects at which the SWE saturates does impose a limitation to the maximum obtainable conversion efficiency in amorphous silicon based solar cells and is still an important drawback for this technology.

The main material properties that could play a role in the SWE are (i) the concentration of impurities, (ii) the hydrogen concentration and its complex bonding structure, and (iii) the disorder in the Si network. From experiments with extremely pure materials, with an oxygen concentration as low as 2×10^{15} cm^{-3}, produced at the Electrotechnical Laboratory (ETL) in Tsukuba, Japan (Kamei et al., 1996), it can be concluded that the light-induced effects are *intrinsic* to the silicon-hydrogen network. Only at impurity concentrations higher than 10^{18} cm^{-3} a correlation can be found between the SWE and the impurity levels (Tsai et al., 1983).

The experimental evidence that the SWE is *intrinsic* to hydrogenated amorphous silicon has made two of the proposed microscopic models plausible. The most favored model has been the hydrogen bond switching model (Stutzmann et al., 1985) which proposes that photoexcited electrons and holes recombine at weak Si-Si bond locations, that the concomitant non-radiative energy release is sufficient to break the bond, and that a back-bonded H atom prevents restoration of the broken bond by a bond switching event. The other plausible model is the charge transfer model (Adler, 1983) that does not invoke H motion and suggests that preexisting spinless centers (positively and negatively charged dangling bonds) are transformed to neutral dangling bonds by capture of excess carriers of opposite sign. *It is noteworthy that both models assume that the defect creation event has a very localized nature.*

Currently, there are experimental results that do not support the local microscopic models while novel characterization techniques are able to detect light-induced changes that affect regions of the Si network that are much more extended. Firstly, pulsed ESR experiments (Yamasaki and Isoya, 1993) show that there is no close spatial correlation (within 0.8 nm) between light-induced dangling bonds and hydrogen as would be the case in the hydrogen bond switching model. A further problem with the H bond switching model is that dangling bonds can in practice be created with the same efficiency at 4.2 K as at 300 K and significant annealing of defects created at 4.2 K occurs at temperatures well below 300 K (Stradins and Fritzsche, 1994), whereas hydrogen is basically immobile at these temperatures, even under intense illumination (Greim et al., 1994). The charge transfer model has also become increasingly unlikely, since the subgap absorption as measured by Photothermal Deflection Spectroscopy (PDS), which is equally sensitive to transitions from charged or neutral states, correlates well with the ESR spin density for stable as well as for light-induced defects (Brandt et al., 1993). The strong correlation does not make it likely that there is a significant contribution of ESR-inactive charged dangling bonds to the SWE.

Increasing evidence suggests that the SWE is sensitive to the hydrogen microstructure and should be detectable as a collective change in the Si network. Hydrogen bonded in configurations leading to a high microstructure parameter R^* (see Section 3.1.1) was shown to give rise to faster metastable defect creation kinetics (Manfredotti et al., 1994, Zafar and Schiff, 1989). Whereas hydrogen in the clustered phase affects the *kinetics* of the SWE, Godet et al., 1996 found that the *saturated density* of metastable defects is correlated with the hydrogen in the dilute phase (i.e. yielding stretching mode vibrations only at 2000 cm^{-1}). Yamasaki and Isoya, 1993 showed that light-induced defects could be found in H-poor regions only, not in hydrogen rich regions.

Accumulating evidence is available now that the SWE manifests itself in changes that are extended over large regions of the silicon network, but perhaps still limited to those regions where hydrogen is dispersed and purely bound as monohydrides up to the solution limit of 2 - 4 % (Daey Ouwens and Schropp, 1996, Acco et al., 1996). The following observations can not be explained by localized bond-switching and charge-transfer events alone.

1. The $\mu\tau$ -product is not a single-valued function of the dangling bond concentration as measured by optical techniques (Han and Fritzsche, 1983).

2. The change in the proton Nuclear Magnetic Resonance (NMR) dipolar spin relaxation time upon light soaking was found to involve 10 % of the 3×10^{21} cm^{-3} bonded hydrogen atoms (Hari et al., 1994). This can only be understood if the structure of the Si network changes in an extended volume since only then the dipolar spin lattice relaxation times of many hydrogen nuclei can be affected.

3. The Si 2p peak detected by core-electron X-ray photoemission spectroscopy (XPS) was found to shift reversibly upon light soaking to a lower binding energy, which implies that not only the Si atoms nearest to the 10^{17} cm^{-3} metastable defects are affected, but also those in an extended region outside of the nearest neighbour shell (Masson et al., 1995).

4. The distinct change in the 1/f noise statistics upon light soaking from non-Gaussian to Gaussian suggests collective long-range interactions rather than locally isolated bond-breaking (Fan and Kakalios, 1994).

5. The ratio of anisotropic to isotropic polarized electroabsorption appears to undergo significant changes *before* the metastable defects appear in CPM measurements on the same sample (Hata et al., 1997). This ratio reflects the bond angle disorder in the silicon network and indicates that dangling bond creation is preceeded by long range network strain.

6. Infrared spectroscopy measurements revealed a reversible significant increase of the monohydride absorption mode after light soaking (Zhao et al., 1995). A possible explanation is that the oscillator strength of this mode has increased due to a structural change of the silicon matrix.

7. Recently it was shown that even the geometrical volume of the material may change due to light soaking (Kong et al., 1997, Kong et al., 1998). The photodilation effect takes place instantaneously when the sample is exposed to light. Although the photostructural origin of the *transient* effect has been called in question (Pashmakov and Fritzsche, 1998), the remaining dilatation *after* extended light exposure (Goto et al., 1998) and the change in the low frequency dielectric constant after the exposure (Yue et al., 1998) are observations that once more suggest that macroscopic changes take place in the Si network.

The observations of large scale structural changes listed above support the viewpoint that long range motion of configurational defects is taking place during light soaking (Fritzsche, 1997). These transporting configurational defects can be mobile dangling bonds, floating bonds (Pantelides, 1986), excited hydrogen in transport states (Santos et al., 1991), or complexes of one or more dangling bonds and hydrogen at a transporting energy level (Biswas et al., 1997, Branz, 1998, Branz, 1998a). The hydrogen collision model by Branz, 1998 is consistent with the spatial separation of light-induced dangling bonds and hydrogen, the observed kinetics under continuous and pulsed light soaking, and the observed macroscopic structural changes, however, the ultimate proof would come from the detection of $\approx 10^{17}$ cm^{-3} metastable complexes of two Si-H bonds.

Recent research targeting at "more stable" materials seems to be successful by addressing the role of hydrogen (e.g., by depositing device-quality material with a very low hydrogen content; Mahan et al., 1991) or the role of the microstructure (e.g., by depositing nanostructured silicon from a plasma close to the powder regime; Roca i Cabarrocas et al., 1998). The latter material comprises crystalline-like regions where no defect creation takes place, probably because configurational defect motion does not take place in these regions, whereas the surrounding material contains more clustered hydrogen in which no large-scale photostructural strain builds up. Previously, these two phases have also been observed in deuterated materials which appeared to be very stable (Sugiyama et al., 1997). The existence of these two phases might also be the origin for the observed near-perfect stability of a-Si:H TFTs made by Hot Wire deposition (Meiling and Schropp, 1997).

Another approach is to avoid amorphous regions altogether and introduce polycrystalline thin films (with either nanometer-scale grains or larger) as the photoabsorber film in photovoltaic devices (Meier et al., 1994, Yamamoto et al., 1998).

5.2 LIGHT-INDUCED CHANGES IN SOLAR CELLS

5.2.1 Single junction cells

Light exposure initially causes a reduction of the conversion efficiency of a-Si:H solar cells. The rate of degradation during continuous illumination at 1 sun intensity is high during the first few tens of hours, but decreases over time such that the cell performance stabilizes after several hundreds of hours. The initial efficiency can be complety recovered by annealing of the device at 150 °C for several hours. The reversibility of the device performance shows that the initial loss is not due to diffusion of ions or dopants, or other irreversible processes such as electromigration. All of the phenomena in solar cells are consistent with metastable defect creation due to the Staebler-Wronski effect discussed in the previous section for individual thin films. A part of the dangling bonds created in the intrinsic part of the p^+-i-n^+ structure acts as recombination centers for the photogenerated carriers. The metastable dangling bonds have an energetic and spatial distribution, which determines their charge state. The space charge distribution of the charged states then modifies the internal electric field profile, which in turn has an additional effect on the carrier collection in the device.

First, we list the most commonly observed effects in p^+-i-n^+ amorphous silicon solar cells.

- The largest relative changes occur in the fill factor (*FF*), the relative changes in the short circuit current J_{sc} are significantly smaller, whereas the open circuit voltage V_{oc} usually does not degrade (see Fig. 5.1).

- Solar cells with a thick intrinsic layer degrade deeper than those with a thin intrinsic layer.

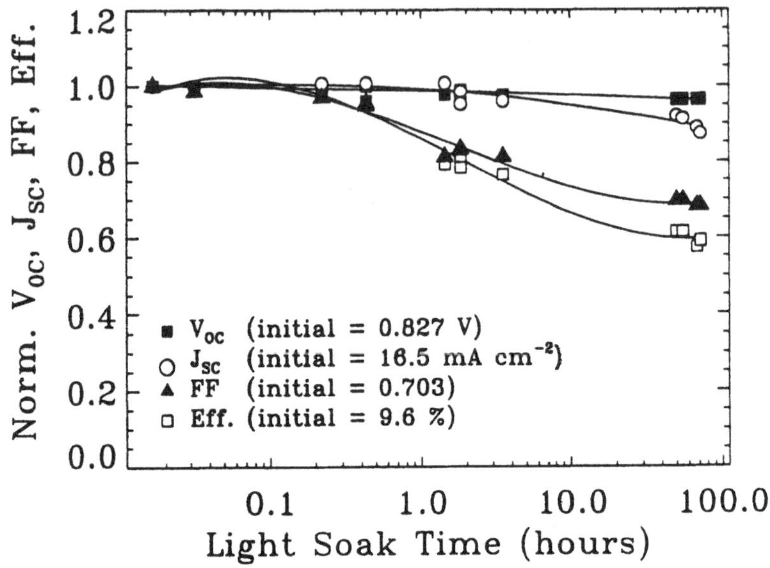

Figure 5.1. Typical relative changes in V_{OC}, J_{SC}, FF, and efficiency during continuous light soaking of a single junction p^+-i-n^+ solar cell with a AM1.5 Global, 100 mW/cm^2 spectrum (Von der Linden, 1994).

- Solar cells with a high impurity concentration (above 10^{18} cm^{-3}) in the intrinsic layer degrade deeper than those with high purity intrinsic layers.
- Solar cells operated at elevated temperatures (60 - 90 °C) stabilize at a higher efficiency than those operated at room temperature or below.
- Cyclic exposure results in stabilization at a higher efficiency than continuous exposure.
- Exposure to high intensity illumination causes deeper degradation than to 1-sun illumination; illumination levels less than 100 mW/cm^2 lead to reduced degradation.

Further, intrinsic layer properties such as the hydrogen microstructure, the void structure, and the Si network order can sometimes be correlated with the device stability (Sugiyama et al., 1997, Roca i Cabarrocas et al., 1998). Frequently, however, the correlation is not clear due to the lack of control over other effects occurring at the p/i and i/n interfaces and within the doped layers. For example, cells with intrinsic layers made from highly hydrogen-diluted silane discharges were shown to stabilize at a higher efficiency than those with an intrinsic layer made from pure silane (Xu et al., 1993, Bennett et al., 1993), but no correlation between these results and the properties of the respective thin film intrinsic materials could be found. A possible explanation for the lack of correlation between the stability of individual intrinsic layers and those incorporated in a p^+-i-n^+ structure can be deduced from the defect pool model for the distribution of defect states in the band gap. The defect pool model (Smith and Wagner, 1988) implies that the dangling bond formation energy depends on the Fermi energy and the energy at which the defect occurs (Bar-Yam and Joannopoulos, 1987). Dangling bonds can be created in three charge states (D^+, D^0, or D^-), corresponding to an occupancy of 0, 1, or 2 electrons. In the intrinsic layer near the p/i interface, where the Fermi level is in the lower half of the band gap, the formation energy of D^+ states is lower than in the intrinsic layer further away from the p/i interface, where the Fermi level is close to mid gap. Consequently, already in the initial state, the intrinsic layer near the p/i interface has a high concentration of positively charged defect states (Brüggeman et al., 1992, Branz and Crandall, 1989). This is illustrated in Chapter 6, Fig. 6.4. It can be expected, that locally created metastable bonds will likewise be created as positively charged defects, thus leading to an enhanced electric field near the p/i interface and a concomitant collapse of the electric field in the bulk (see Fig. 5.1). This limits the collection of free charge carriers from the body of the cell, where, in addition, light-induced neutral dangling bonds act as recombination centers. This model further shows that the Staebler-Wronski effect can be regarded as a bulk effect, while the local properties of the material are sensitive to band bending effects near interfaces.

In order to reduce the SWE related degradation of *solar cells*, many different methods have been investigated to circumvent the effects by *device design* rather than attempting to create the ultimately stable amorphous network. Most approaches address the question how the *degraded* electric field profile of the p^+-i-n^+ device can be optimized.

- An engineering solution is to make the intrinsic absorber material as thin as possible in order to maintain a high electric field after degradation. At the same time, the absorbed fraction of the incident light is maximized by utilizing optical light confinement techniques made possible by textured electrodes and enhanced multilayer back reflectors (see Chapter 4).

- Band-gap profiling of the i-layer can assist the transport of the minority carriers (holes) in the low field region that is brought about by light-induced degradation. This was suggested, among others, by Dalal and Baldwin,

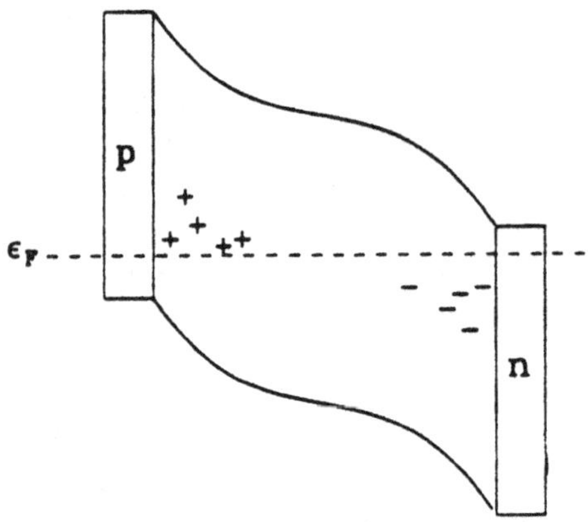

Figure 5.2 Schematic band diagram of a p$^+$-i-n$^+$ solar cell after light soaking. The electric field distribution is distorted due to positively charged defects near the p/i interface and negatively charged defects near the i/n interface.

1993. The schemes employed in a-SiGe:H alloy cells to enhance the initial efficiency (Guha et al., 1989), may well be adapted to enhance the stabilized efficiency. No systematic study on band-gap grading for stabilized efficiency optimization has been made public, but many laboratories have implemented profiling schemes derived from empirical tests.

- Field redistribution using graded low-level impurity doping has been demonstrated by Fischer et al., 1993, but the defects associated with impurities prevent a high value of the stabilized efficiency.
- A method of mobility grading by stacking layers with different carrier mobilities has been proposed by Von Roedern, 1992.
- A method of latent charged defect engineering in order to manipulate the internal field profile in the degraded state was demonstrated by Schropp et al., 1993.

The highest stabilized efficiencies for single junction cells reported so far are 8.8 % (Hishikawa et al., 1994), for p$^+$-i-n$^+$ cells, and 9.3 % (Guha et al., 1998) for n$^+$-i-p$^+$ cells. Table 4.2 gives a more complete listing.

5.2.2 Multijunction cells

Single junction cells with a very thin intrinsic absorber layer (< 250 nm) have very good stability, but even the most advanced techniques for optical confinement are not sufficient to improve the stabilized efficiency beyond that of slightly thicker cells. Therefore, multijunction cells, without altering band gaps, are presently used to absorb more light, while each component junction has a sufficiently high internal electric field to assure good collection even in the degraded state.

The main reason for enhanced stability of multijunction cells is that the electrical field can be maintained in the thin component cells even after the concentration of metastable defects has been driven to saturation. The reduced photogenerated carrier density in the bottom cell due to the filter effect of the top cell is a much less significant reason for the enhanced stability, because the saturated defect density is only weakly dependent on the optical irradiation density (Gleskova et al., 1993). Consistently, no increased degradation is observable in single junction cells when the generation rate is increased by utilizing optical confinement.

Nevertheless, the filter effect of the top cell induces a generation profile in the bottom cell that is different from that in a single junction cell. The generation profile in turn influences the electric field distribution. Furthermore, in two-

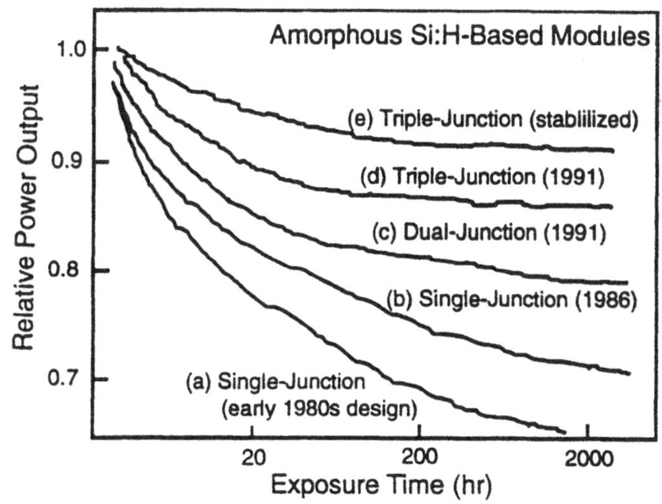

Figure 5.3. Normalized solar cell output for single junction and multijunction devices illustrating the enhanced stability for multijunction devices and the positive effect of other design improvements over the years (reprinted from Kazmerski, 1997, with permission from Elsevier Science).

terminal multijunction cells, the voltage bias is distributed over the junctions in a manner that depends on the load resistance and the spectrum of the incident light. As the values of the fill factor and magnitudes of the short circuit current of the component cells may degrade at different rates, it may not always be the same cell that dictates the overall fill factor, and current matching may not always be maintained during the operational life of a multijunction cell.

Catalano, 1990 modeled the expected performance of various multijunction solar cells and found that the measured data of experimental devices agreed remarkably well with the calculated behaviour. In particular, when multijunction cells contain different absorber layers with differing band gaps, the lower stability of a-SiGe:H component cells (Xu et al., 1993) compared to a-Si:H cells has to be taken into account in order to correctly predict the stabilized performance. On the other hand, in some cases the thin top cell may degrade deeper as it has a larger probability to suffer from light-induced shunting effects (McMahon and Bennet, 1992).

Any current mismatch will lead to charge accumulation at the internal tunnel-recombination junction(s). This changes the voltage bias distribution over the component cells and thus alters the rate at which these cells degrade. The design of the most stable multibandgap solar cell at present is such that the performance is "top cell limited", i.e. the current is dictated by the cell that does not contain an a-SiGe:H absorber (Yang and Guha, 1992). Recently, Kazmerski, 1997 summarized (Fig. 5.3) the progress achieved in the stability of amorphous silicon based modules, among others, due to the development of the multijunction cell structure and the hydrogen-dilution technique for the intrinsic absorber layers.

References

Acco, S., D.L. Williamson, P.A. Stolk, F.W. Saris, M.J. van den Boogaard, W.C. Sinke, W.F. van der Weg, and S. Roorda, *Hydrogen solubility and network stability in amorphous silicon*, Phys. Rev. B **53** (1996) 4415-4427.

Adler, D., *Origin of the photo-induced changes in hydrogenated amorphous silicon*, Solar Cells **9** (1983) 133-148.

Bar-Yam, Y., and J.D. Joannopoulos, *Theories of defects in amorphous semiconductors*, J. Non-Cryst. Solids **97 & 98** (1987) 467-474.

Bennett, M., K. Rajan, and K. Kritikson, *Amorphous silicon based solar cells deposited from H_2-diluted SiH_4 at low temperatures*, Proc. of the 23rd IEEE PV Spec. Conf. (1993) 845-849.

Biswas, R., Q. Li, B.C. Pan, and Y. Yoon, *Reactivity and migration of hydrogen in a-Si:H*, in: Amorphous and Microcrystalline Silicon Technology - 1997, edited by S. Wagner, M. Hack, E.A. Schiff, R. Schropp, and I. Shimizu, Materials Research Society Symp. Proc. **467** (1997) 135-140.

Brandt, M.S., A. Asano, and M. Stutzmann, *Are there charged dangling bonds in device quality amorphous silicon?*, in: Amorphous Silicon Technology - 1993, edited by E.A. Schiff, M.J. Thompson, A. Madan, K. Tanaka, P.G. LeComber, Materials Research Society Symp. Proc. **297** (1993) 201-206.

Branz, H.M., and R.S. Crandall, *Defect equilibrium thermodynamics in hydrogenated amorphous silicon: consequences for solar cells*, Solar Cells **27** (1989) 159-168.

Branz, H.M., *Hydrogen collision model of light-induced metastability in hydrogenated amorphous silicon*, Solid State Commun. **105** (1998) 387-391.

Branz, H.M., *Hydrogen collision model of the Staebler-Wronski effect: microscopics and kinetics*, in: Amorphous and Microcrystalline Silicon Technology - 1998, edited by R. Schropp, H. Branz, S. Wagner, M. Hack, and I. Shimizu, Materials Research Society Symp. Proc. **507** (1998) in print.

Brüggeman, R., C. Main, and G.H. Bauer, *Simulation of the steady state and transient phenomena in a-Si:H pin structures and films*, in: Amorphous Silicon Technology - 1992, edited by M.J. Thompson, Y. Hamakawa, P.G. LeComber, A. Madan, and E. Schiff, Materials Research Society Symp. Proc. **258** (1992) 729-734.

Catalano, A., *Advances in a-Si:H alloys for high efficiency devices*, Proc. 21st IEEE Photovoltaic Specialists Conference (IEEE, New York, 1990) 36-40.

Daey Ouwens, J., and R.E.I. Schropp *Hydrogen microstructure in hydrogenated amorphous silicon*, Phys. Rev. B **54** (1996) 17759-17762.

Dalal, V.L., and G. Baldwin, *Design and fabrication of graded bandgap solar cells in amorphous Si and alloys*, in: Amorphous Silicon Technology - 1993, edited by E.A. Schiff, M.J. Thompson, A. Madan, K. Tanaka, P.G. LeComber, Materials Research Society Symp. Proc. **297** (1993) 833-838.

Fan, J., and J. Kakalios, *Light-induced changes of the non-Gaussian 1/f noise statistics in doped hydrogenated amorphous silicon*, Phil. Mag. B **69** (1994) 595-608.

Fischer, D., N. Wyrsch, C.M. Fortmann, and A.V. Shah, *Amorphous silicon solar cells with graded low-level doped i-layers characterised by bifacial measurements*, Proc. of the 23rd IEEE PV Spec. Conf. (1993) 878-884.

Fritzsche, H., *Search for explaining the Staebler-Wronski effect* in: Amorphous and Microcrystalline Silicon Technology - 1997, edited by S. Wagner, M. Hack, E.A. Schiff, R. Schropp, and I. Shimizu, Materials Research Society Symp. Proc. **467** (1997) 19-30.

Gleskova, H., J.N. Bullock, and S. Wagner, *Isolating the rate of light-induced annealing of the dangling-bond defects in a-Si:H*, J. Non-Cryst. Solids **164-166** (1993) 183-186.

Godet, C., P. Morin, and P. Roca i Cabarrocas, *Influence of the dilute-phase SiH bond concentration on the steady-state defect density in a-Si:H*, J. Non-Cryst. Solids **198-200** (1996) 449-452.

Goto, T., S. Nonomura, M. Nishio, N. Masui, S. Nitta, M. Kondo, and A. Matsuda, *Detection of photoinduced structural change in a-Si:H by bending effect*, J. Non-Cryst. Solids **227-230** (1998) 263-266.

Greim, O., J. Weber, Y. Baer, and U. Krol, *Hydrogen diffusion in a-Si:H stimulated by intense illumination*, Phys. Rev. B **50** (1994) 10644-10648.

Guha, S., J. Yang, A. Pawlikiewicz, T. Glatfelter, R. Ross, and S.R. Ovshinsky, *Band-gap profiling for improving the efficiency of amorphous silicon alloy solar cells*, Appl. Phys. Lett. **54** (1989) 2330-2332.

Han, D., and H. Fritzsche, *Study of light-induced creation of defects in a-Si:H by means of single and dual-beam photoconductivity*, J. Non-Cryst. Solids **59 & 60** (1983) 397-400.

Hari, P., P. Taylor, and R.A. Street, *Effect of light soaking on the local motion of hydrogen in hydrogenated amorphous silicon*, in: Amorphous Silicon Technology - 1994, edited by E.A. Schiff, M. Hack, A. Madan, M. Powell, and A. Matsuda, Materials Research Society Symp. Proc. **336** (1994) 329-334.

Hata, N., T. Kamei, H. Okamoto, and A. Matsuda, *Polarized electroabsorption in pulse and continuous light-soaked a-Si:H - Structural change other than defect creation*, in: Amorphous and Microcrystalline Silicon Technology - 1997, edited by S. Wagner, M. Hack, E.A. Schiff, R. Schropp, and I. Shimizu, Materials Research Society Symp. Proc. **467** (1997) 61-66.

Hishikawa, Y., K. Ninomiya, E. Maruyama, S. Kuroda, A. Terakawa, K. Sayama, H. Tarui, M. Sasaki, S. Tsuda, and S. Nakano, *Approaches for stable multi-junction a-Si solar cells*, 1st World Conference on Photovoltaic Energy Conversion, (Proc. 24th IEEE PV Specialists Conference, Waikoloa, HI, USA, December 1994) 386-393.

Kamei, T., N. Hata, A. Matsuda, T. Uchimya, S. Amano, K. Tsukamoto, Y. Yoshioka, and T. Hirao, *Deposition and extensive light soaking of highly pure hydrogenated amorphous silicon*, Appl. Phys. Lett. **68** (1996) 2380-2382.

Kazmerski, L.L., *Photovoltaics: A review of cell and module efficiencies* Renewable and Sustainable Energy Reviews **1**, Nos. 1/2 (1997) 71-170.

Kong, G., D. Zhang, G. Yue, and X. Liao, *Photodilatation effect in undoped a-Si:H films*, Phys. Rev. Lett. **79** (1997) 4210-4213.

Kong, G., *Light excited structural instability in a-Si:H*, in: Amorphous and Microcrystalline Silicon Technology - 1998, edited by R. Schropp, H. Branz, S. Wagner,

M. Hack, and I. Shimizu, Materials Research Society Symp. Proc. **507** (1998) in print.

Mahan, A.H., J. Carapella, B.P. Nelson, R.S. Crandall, and I. Balberg, *Deposition of device quality, low H content amorphous silicon*, J. Appl. Phys. **69** (1991) 6728-6730.

Manfredotti, C., F. Fizzotti, M. Boero, P. Pastorino, P. Polesello, and E. Vittone, *Influence of hydrogen-bonding configurations on the physical properties of hydrogenated amorphous silicon*, Phys. Rev. B **50** (1994) 18046-18053.

Masson, D.P., A. Ouhlal, and A. Yelon, *Long-range structural relaxation in the Staebler-Wronski effect*, J. Non-Cryst. Solids **190** (1995) 151-156.

McMahon, T.J., and M.S. Bennet, *Film morphology, excess shunt current and stability in triple-junction cells*, in: Amorphous Silicon Technology - 1992, edited by M.J. Thompson, Y. Hamakawa, P.G. LeComber, A. Madan, and E. Schiff, Materials Research Society Symp. Proc. **258** (1992) 941-946.

Meier, J., R. Flückiger, H. Keppner, and A. Shah, *Complete microcrystalline p-i-n solar cell - crystalline or amorphous cell behavior*, Appl. Phys. Lett. **65** (1994) 860-862.

Meiling, H., and R.E.I. Schropp, *Stable amorphous silicon thin film transistors*, Appl. Phys. Lett. **70** (1997) 2681-2683.

Pantelides, S.T., *Defects in amorphous silicon: a new perspective*, Phys. Rev. Let. **57** (1986) 2979-2982.

Pashmakov, B., and H. Fritzsche, *Comment on "Photodilatation effect in undoped a-Si:H films*, Phys. Rev. Lett. **80** (1998) 5704-5705.

Roca i Cabarrocas, P., *Plasma deposition of silicon clusters: a means to produce medium range ordered silicon thin films*, in: Amorphous and Microcrystalline Silicon Technology - 1998, edited by R. Schropp, H. Branz, S. Wagner, M. Hack, and I. Shimizu, Materials Research Society Symp. Proc. **507** (1998) in print.

Santos, P.V., N.M. Johnson, and R.A. Street, *Light-enhanced hydrogen motion in a-Si:H*, Phys. Rev. Let. **67** (1991) 2686-2689.

Schropp, R.E.I., A. Sluiter, M.B. von der Linden, and J. Daey Ouwens, *Stability of amorphous silicon materials incorporated in solar cells and intrinsic layer profiling for enhanced stabilized performance*, J. Non-Cryst. Solids **164-166** (1993) 709-712.

Smith, Z.E., and S. Wagner, *Implications of the 'Defect Pool' concept for 'metastable' and 'stable' defects in amorphous silicon*, in: "Amorphous Silicon and Related Materials", edited by H. Fritzsche (World Scientific, Singapore, 1988) 409-460.

Staebler, D.L., and C.R. Wronski, *Reversible conductivity changes in discharge produced amorphous Si*, Appl. Phys. Lett. **31** (1977) 292-294.

Staebler, D.L., and C.R. Wronski, *Optically induced conductivity changes in discharge-produced hydrogenated amorphous silicon*, J. Appl. Phys. **51** (1980) 3262-3268.

Stradins, P., and H. Fritzsche, *Photoinduced creation of metastable defects in a-Si:H at low temperatures and their effect on the photoconductivity*, Phil. Mag. B **69** (1994) 121-139.

Stutzmann, M., W.B. Jackson, and C.C. Tsai, *Light-induced metastable defects in hydrogenated amorphous silicon: a systematic study*, Phys. Rev. B **32** (1985) 23-47.

Sugiyama, S., J. Yang, and S. Guha, *Improved stability against light-exposure in deuterated amorphous silicon alloy solar cells*, in: Amorphous and Microcrystalline Silicon Technology - 1997, edited by S. Wagner, M. Hack, E.A. Schiff, R. Schropp, and I. Shimizu, Materials Research Society Symp. Proc. **467** (1997) 49-54.

Tsai, C.C., J.C. Knights, R.A. Lujan, B. Wacker, B.L. Stafford, and M.J. Thompson, *Amorphous Si prepared in a UHV plasma deposition system*, J. Non-Cryst. Solids **59 & 60** (1983) 731-734.

Von der Linden, M.B. *Electronic Defects in Amorphous Silicon. Materials and Devices*, Ph.D. Thesis, Utrecht University, The Netherlands (1994).

Von Roedern, B., *Innovative optimization procedures for solar cells based on a unique model for junction optimization*, 12th International E.C. Photovoltaic Solar Energy Conference 1994, Eds. R. Hill, W. Palz, and P. Helm (Stephens and Associates, 1994) 1354-1358.

Xu, X., J. Yang, and S. Guha, *On the lack of correlation between film properties and solar cell performance of amorphous silicon-germanium alloys*, Appl. Phys. Lett. **62** (1993) 1399-1401.

Yamamoto, K., *Thin film poly-Si solar cell on glass substrate fabricated at low temperature*, in: Amorphous and Microcrystalline Silicon Technology - 1998, edited by R. Schropp, H. Branz, S. Wagner, M. Hack, and I. Shimizu, Materials Research Society Symp. Proc. **507** (1998) in print.

Yamasaki, S., and J. Isoya, *Pulsed ESR study of light-induced metastable defect in a-Si:H*, J. Non-Cryst. Solids **164-166** (1993) 169-174.

Yang, J., and S. Guha, *Double-junction amorphous silicon based solar cells with 11 % stable efficiency*, Appl. Phys. Lett. **61** (1992) 2917-2919.

Yue, G., G. Kong, D. Zhang, Z. Ma, S. Sheng, and X. Liao, *Dielectric response and its light-induced change in undoped a-Si:H films below 13 MHz*, Phys. Rev. B **57** (1998) 2387-2392.

Zafar, S., and E.A. Schiff, *Hydrogen-mediated model for defect metastability in hydrogenated amorphous silicon*, Phys. Rev. B **40** (1989) 5235-5238.

Zhao, Y., D. Zhang, G. Kong, G. Pan, and X. Liao, *Evidence for light-induced increase of Si-H bonds in undoped a-Si:H*, Phys. Rev. Lett. **74** (1995) 558-561.

Amorphous and Microcrystalline Silicon Solar Cells: Modeling, Materials and Device Technology

II Modeling of Amorphous Silicon Solar Cells

6 ELECTRICAL DEVICE MODELING

Nature has that many parameters...

—Stephen Fonash, April 15, 1998

6.1 INTRODUCTION

The simulation of the electrical and optical behavior of semiconductor devices has been established as an essential tool for both the improvement of existing devices and for the development of new ones. There is no doubt that the role of device modeling will increase in the future. Device modeling involves the numerical solution of a set of equations, which form a mathematical model for device operation, together with models that describe the material properties. A number of sophisticated device program packages are already commercially available on the market such as Medici, 1997 from TMA company or Atlas from

The authors wish to thank Stephen Fonash from PennState University and Jože Furlan from the University of Ljubljana who in a review of this chapter offered many helpful suggestions.

SILVACO company. These programs are mostly designed for modeling a broad range of crystalline semiconductor devices. At present they are continuously updated to offer possibilities to model also polysilicon and amorphous silicon based devices such as thin film transistors (TFTs) and solar cells. The advantage of these programs is that they are modular, so the users need to acquire only the minimum set of modules to meet their needs. A comprehensive set of two dimensional (2-D) device simulation tools and common libraries such as parser, grid generation algorithms, solvers and external interfaces are integrated in these programs. Still, the models that are used in these packages to describe material properties of amorphous silicon are relatively simple. Therefore, several groups have developed their own computer programs for modeling amorphous and microcrystalline silicon solar cells, in which more flexible and sophisticated models describing amorphous silicon electronic properties are implemented. Examples are: the model of Hack and Shur, 1985, the program AMPS developed at PennState University (McElheny et al., 1988), the ASPIN program from Ljubljana University (Smole and Furlan, 1992), and the ASA package developed at Delft University of Technology (Zeman et al., 1997).

An important and exciting part of computer modeling is the development of new models that describe material properties and device operation processes more accurately. The models form a general mathematical description of device behavior under specific conditions. Only under the most simple conditions the general mathematical description can be approximated by analytical solutions. In most cases the solution of a set of mathematical equations (often called model equations), is carried out numerically by a computer program. This kind of computer programs are also referred to as device simulation programs or device simulators. A set of input parameters is required by the computer program and as a result it can deliver: i) external properties of a layer or device that are measurable quantities, ii) internal properties of a layer or device that can not be measured directly or not at all. In the case of solar cells the external properties are the dark and illuminated J-V characteristics or the spectral response. The internal properties are the electric field, the concentration of free and trapped charge carriers, the space charge or the recombination rate as a function of the position in the solar cell.

Accurate calibration of model parameters, which is the assignment of proper values to the input parameters, is an important step in modeling. The calibration procedure is based on comparing the simulated external properties with experimental data. When the calibrated computer model reproduces a broad range of experimental results one can be confident that the models and the values of the model parameters are correct and can start using the predictive power of such a computer model. Then, one can design test structures for the experimental extraction of device parameters and optimize the solar cell structure for the best performance. When no agreement between the simulated and measured data can be reached, the models of the material properties or the physical processes in the device are probably incorrect and need to be adapted. This procedure of model development, model calibration, prediction, and deviation of the model from the experiment contributes to a better knowledge of the material properties and the physics controlling the device operation.

Although the available programs still have uncertainties in the values of several model parameters, their use already offers some notable advantages over technological development. Using simulations it is possible to examine the influence of model parameters, which cannot be determined experimentally, or they can be set independently from each other, so that the impact of small changes in a device configuration can be determined much faster and more reliable. Moreover, as computer hardware components are becoming less expensive and more powerful, simulations are getting cheaper in contrast to experimental equipment which is becoming increasingly costly.

Most of the existing simulation programs are designed for crystalline semiconductor devices. These programs are based on the solution of the semiconductor equations (Kurata, 1982, Selberherr, 1984), and on the physical models that describe the semiconductor material properties. This approach can also be used in modeling a-Si:H based devices. However, in the case of a-Si:H, special attention has to be paid to modeling the continuous density of states (DOS) distribution in the band gap and the recombination-generation (R-G) statistics of these states.

Modeling of a-Si:H solar cells started in the early 80s and since then it has made large progress. Different research groups began to carry out computer analysis of a-Si:H solar cells using simplified or approximated semiconductor equations (Okamoto et al., 1982, Crandall, 1983, Sichanugrist et al., 1984). Over the years several advanced computer programs have been developed for modeling a-Si:H solar cells (Swartz, 1982, Swartz, 1984, Chen and Lee, 1982, Hack and Shur, 1985, Ikegaki et al., 1985, McElheny et al., 1988, Tasaki et al., 1988, Wentinck, 1988, Smole and Furlan, 1992, Block, 1993, Stiebig et al., 1994). These computer programs differ in the choice of the independent variables in the semiconductor equations, in the numerical solution techniques, in the possibilities to describe material properties as a function of the position in the device, in the description of the DOS distribution in the band gap of a-Si:H based materials, in the R-G statistics of the localized states, and in the contact treatment. Some of the groups have implemented in their programs special models for the interface behavior between adjacent layers in the solar cells, such as transparent conductive oxide (TCO) and the a-Si:H based layer (Smole et al., 1994) or between the doped a-Si:H layers that form a tunnel-recombination junction in multijunction solar cells (Hou et al., 1991, Willemen et al., 1994).

All above-mentioned computer programs are limited to one dimensional (1-D) modeling. In principle 1-D modeling is well suited for a-Si:H solar cells on flat substrates. However, there are at least two important developments in the field of a-Si:H solar cells that point out that two dimensional (2-D) modeling is required in the future for more accurate modeling. The first development is the use of textured substrates in all of the present high-efficiency a-Si:H solar cells, which introduces spatial variations in the device structure. The second development in thin film solar cells is the application of microcrystalline silicon, which is not a spatially homogeneous material.

At present, the main effort in the electrical modeling of a-Si:H solar cells is aiming at two goals. First is improving, extending, and testing of the existing computer models which have to be constantly updated to reflect a more profound knowledge of a-Si:H based materials and processes in an increasingly complex structure of the solar cells. Second is establishing reliable values of physical device model parameters. Most of the model parameters are still not known precisely and this represents a considerable limitation to accurate modeling of a-Si:H based solar cells. Because of the steady-state character of power conversion in solar cells, modeling of their performance is restricted to time-independent situations. Modeling of time-dependent processes such as time-of-flight (TOF) (Tiedje, 1984) or transient photocurrent in a-Si:H structures provides an extra possibility to determine and extract carrier trapping and recombination parameters of a-Si:H.

6.2 SPECIFIC ISSUES IN AMORPHOUS SILICON DEVICE MODELING

In a-Si:H, the similarity of the short range order of atomic configurations to that in crystalline silicon is responsible for its semiconductor properties. The impact of the lack of long range order in the atomic structure of a-Si:H on its electronic properties is best reflected in the energy band diagram of the DOS distribution. The density of states in crystalline silicon is represented by a parabolic distribution near the top of the valence band and the bottom of the conduction band. There is a well-defined band gap between the valence and conduction bands. Allowed states within the band gap are introduced by the incorporation of acceptor- or donor-type impurities, deep-level impurities, and by defects in the semiconductor lattice.

Due to the spatial disorder in the atomic structure of a-Si:H the periodicity of the potential energy, normally associated with the lattice in crystalline silicon, is disturbed. This leads to broadening of the conduction and valence bands into the so called tails, which extend into the band gap. Structural defects are present in the disordered network of a-Si:H, which give rise to electron states around the center of the band gap. Therefore, a-Si:H exhibits a continuous density of states in the band gap and there are no well-defined conduction and valence band edges. The models for the DOS distribution in a-Si:H are presented in Section 6.4.

The spatial disorder in a-Si:H not only leads to broadening of the conduction- and valence-band states into the tails but also to localization of these states. Therefore, the continuous DOS distribution is characterized by three different regions: (i) extended states above the mobility edge of the conduction band, (ii) extended states below the mobility edge of the valence band, (iii) localized states between the mobility edges. The mobility edges define the mobility gap in a-Si:H. The mobility of charge carriers in the localized states is much less than the mobility of carriers in the extended states.

Due to the continuous DOS distribution the transport in amorphous silicon is much more complex than in its crystalline counterpart. In crystalline semiconductors the electric current is carried by the motion of electrons in extended states of the conduction band and holes in the extended states of the valence band. In a-Si:H the electric current is carried not only in extended states of the conduction and valence band, but also the states in the mobility gap are involved in the transport. There are three basic transport mechanisms in amorphous semiconductors: conventional extended-state transport, gap-state multiple trapping, and hopping.

In the multiple trapping mechanism an electron moves in an electric field only when it is in the extended states above the mobility edge. It can be captured by a localized band-tail state and stays immobile in this state until it is re-emitted back into the band of extended states. The same process applies for holes. The hopping transport mechanism involves the localized states in the mobility gap. The electrons or holes tunnel from one localized state to another localized state and this process is thermally activated. In this case the energy of the carriers is always lower than the energy corresponding to the mobility gap. It is assumed (Mott and Davis, 1979) that at room temperature the multiple trapping mechanism dominates the transport in amorphous semiconductors, while the hopping transport mechanism is dominant at low temperatures. Since the operational temperature of solar cells is such that hopping transport can be neglected, it is not included in the mathematical model for a-Si:H device operation. The transport with the multiple trapping process can be described by the concentration of carriers in the extended states and their mobility in the extended states. In this way it does not differ from the description of the transport process in crystalline silicon. This similarity allows us to use the same set of basic semiconductor equations, which form a mathematical model for device operation in crystalline silicon, also for a-Si:H based devices.

Due to the low density of states in the band gap of crystalline semiconductors modeling of the R-G rate and the space charge is greatly simplified. The R-G rate is often successfully calculated using the effective electron and hole lifetimes in the Shockley-Read-Hall (SRH) formula (Shockley and Read, 1952). The contribution of carriers, which are trapped in the states in the band gap, to the electric charge is usually neglected. We refer to this approach as the lifetime model of R-G statistics (S. Fonash and H. Zhu, 1998). In the early computer models of a-Si:H solar cells this model was widely used (Swartz, 1982, Ikegaki et al., 1985). However, the continuous distribution of the DOS in the band gap of a-Si:H strongly affects the trapping and recombination processes and the trapped charge in the localized states thus can not be ignored. Therefore, gradually more sophisticated models of the DOS distribution in the band gap of a-Si:H emerged and were implemented in the computer programs (Chen and Lee, 1982, Swartz, 1984, Hack and Shur, 1985, McElheny et al., 1988, Zeman et al., 1995). We describe this approach, which models the recombination and trapping by using a quasi-continuous DOS distribution in the band gap with appropriate R-G statistics, as the DOS model (S. Fonash and H. Zhu, 1998). The DOS model reflects reality more accurately but on the other hand one has to deal with a large number of additional input parameters.

A new requirement in modeling of a-Si:H solar cells emerged when the wide band gap p-type a-SiC:H was implemented in the solar cell (Section 4.2.2). It became clear that an adequate computer model for a-Si:H solar cells had to deal with the physics of heterojunction devices. Tasaki et al., 1988 were the first to report on a computer model that allowed to work with discontinuities in semiconductor properties at the heterojunctions and with graded materials. Furthermore, the multijunction concept of a-Si:H based solar cells, which uses tunneling assisted recombination at the interface between two adjacent junctions, has pushed for the development of models, that could describe the tunnel-recombination processes at this interface. This interface is described as the tunnel-recombination junction (TRJ) in the literature. A computer model that intends to simulate a tandem or triple a-Si:H solar cell as a complete device has to contain a model for TRJs. Two groups have reported on the development of models for the TRJ, the group from PennState University (Hou et al., 1991) and from Delft University of Technology (Willemen et al., 1994).

The introduction of textured substrates for a-Si:H solar cells, the application of microcrystalline silicon into solar cells, and a recent new design of a-Si:H transverse junction solar cell (Kroon et al., 1998) are the main stimulating forces to extend 1-D modeling of a-Si:H based solar cells to, at least, 2-D modeling. This process is under way and one can notice that this is happening in two ways. The companies such as TMA and SILVACO, which offer robust 2-D device simulators designed for crystalline semiconductor devices, have been implementing more or less sophisticated DOS models for R-G statistics into their packages in order to simulate a-Si:H and polysilicon devices properly. On the other hand, some of the groups that have developed their own 1-D computer programs for modeling a-Si:H devices are extending them into 2-D (Sawada et al., 1993, Fantoni et al., 1996, Zimmer et al., 1998). However, a 2-D device simulator that contains a general and flexible DOS model as well as a model for TRJ is at present still not available.

In modeling of crystalline semiconductor devices, activities are largely concentrating on describing the carrier lifetimes and mobilities as functions of e.g. temperature, doping concentrations, and electric field. The exact quantitative descriptions of these dependencies for a-Si:H are not available at present.

6.3 SET OF DEVICE MODEL EQUATIONS

6.3.1 *Basic set of semiconductor equations*

The basic set of semiconductor equations represents a mathematical description of semiconductor device operation under non-equilibrium conditions. A computer program solves this set of equations together with the imposed boundary conditions. The basic semiconductor equations include Poisson's equation (6.1), the continuity equations for electrons (6.2) and holes (6.3) and the equations for electron (6.4) and hole (6.5) current densities. In the absence of a magnetic field and for a uniform temperature in the device the equations have the following form

$$\mathrm{div}\,(\varepsilon\ \mathrm{grad}\psi) = -\rho \tag{6.1}$$

$$\frac{\partial n}{\partial t} = \frac{1}{q}\mathrm{div}J_n + G_{opt} - R_{net} \tag{6.2}$$

$$\frac{\partial p}{\partial t} = -\frac{1}{q}\mathrm{div}J_p + G_{opt} - R_{net} \tag{6.3}$$

$$J_n = n\mu_n\ \mathrm{grad}E_{\mathrm{FN}} \tag{6.4}$$

$$J_p = -p\mu_p\ \mathrm{grad}E_{\mathrm{FP}} \tag{6.5}$$

In these equations ε is the permittivity of the semiconductor, ψ is the electrostatic potential with reference to the vacuum level $\mathrm{E_0}$, ρ is the space charge density, J_n (J_p) is the electron (hole) current density, G_{opt} is the optical generation rate and R_{net} denotes the net recombination-generation rate of electrons and holes. n (p) is the free electron (hole) concentration, μ_n (μ_p) is the electron (hole) mobility, E_{FN} (E_{FP}) is the electron (hole) quasi-Fermi energy level and $\partial n/\partial t$ ($\partial p/\partial t$) is time rate change in the electron (hole) concentration.

6.3.2 Band diagram

The a-Si:H based solar cell is a typical example of a heterojunction device, as it incorporates one or more junctions between two dissimilar semiconductor materials. This fact has serious implications for a computer model which has to take into account the different properties of separate layers in device. Fig. 6.1 shows the energy band diagram under equilibrium conditions for a typical single junction a-Si:H solar cell.

In general, E_{C} represents the minimum energy of the conduction band (CB). In a-Si:H based material this energy is related to the mobility edge of the conduction band. This energy level can be related to the vacuum energy level E_0 by:

$$E_{\mathrm{C}} = E_0 - q\psi - \chi \tag{6.6}$$

where χ denotes the electron affinity. E_{V} is the maximum energy level of the valence band (VB). In a-Si:H based material this energy is related to the mobility edge of the valence band. This energy level is related to the vacuum energy level E_0 by

$$E_{\mathrm{V}} = E_0 - q\psi - \chi - E_{\mathrm{G}} \tag{6.7}$$

where E_{G} denotes the band gap energy; in case of a-Si:H based materials E_{G} represents the mobility gap E_{gap}^{mob}. The terms mentioned above are explained in Section 6.5.

The concentration of free electrons and holes in thermal equilibrium is determined by the position of the Fermi level E_{F} in the device and is calculated from the charge neutrality principle using Fermi-Dirac or Maxwell-Boltzmann statistics. Under non-equilibrium conditions, the concentration of free carriers is described by the quasi-Fermi energy level for electrons E_{FN} and the quasi-Fermi energy level for holes E_{FP}. In thermal equilibrium

$$E_{\mathrm{F}} = E_{\mathrm{FN}} = E_{\mathrm{FP}} \tag{6.8}$$

6.3.3 Carrier statistics

The electron and hole concentrations in semiconductors are defined by the Fermi-Dirac distributions and a parabolic density of states (Pierret, 1987). For the range of operation of most semiconductor devices the Maxwell-Boltzmann

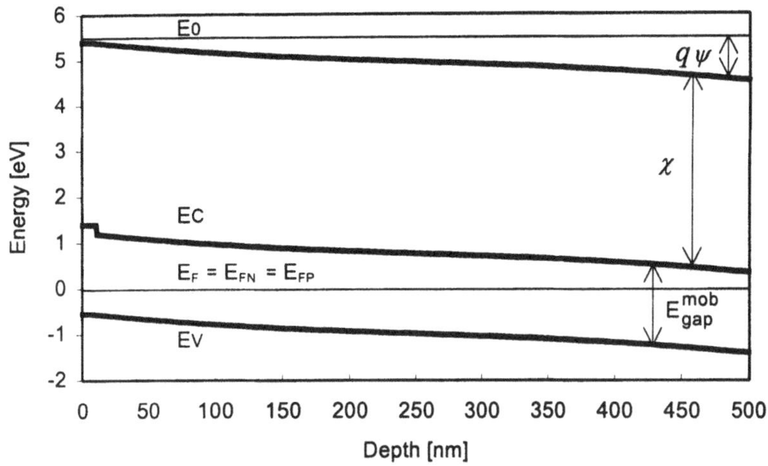

Figure 6.1. The energy band diagram of a typical single junction a-Si:H solar cell under equilibrium conditions.

statistics for the occupation of free electron and hole states can be used. This is the case when the distance between the quasi-Fermi levels and the corresponding band-gap edges is more than 2kT. The free carrier densities are then approximated by

$$n = N_C \exp\left(\frac{E_{FN}-E_C}{kT}\right) = N_C \exp\left(\frac{E_{FN}-E_0+q\psi+\chi}{kT}\right) \quad (6.9a)$$

$$p = N_V \exp\left(\frac{E_V-E_{FP}}{kT}\right) = N_V \exp\left(\frac{E_0-q\psi-\chi-E_G-E_{FP}}{kT}\right) \quad (6.9b)$$

where N_C (N_V) is the effective density of states in the conduction (valence) band. From equations (6.9a) and (6.9b) follows

$$E_{FN} = E_C + kT\ln\left(\frac{n}{N_C}\right) = E_0 - q\psi - \chi + kT\ln\left(\frac{n}{N_C}\right) \quad (6.10a)$$

$$E_{FP} = E_V - kT\ln\left(\frac{p}{N_V}\right) = E_0 - q\psi - \chi - E_G - kT\ln\left(\frac{p}{N_V}\right) \quad (6.10b)$$

6.3.4 Set of model equations

Power conversion in solar cells is considered to be steady state operation. The structure and dimensions of a-Si:H based solar cells allow to restrict the model to one dimension (1-D). Under steady state conditions, using expressions (6.10a) and (6.10b) in equations (6.4) and (6.5), respectively, and after a bit of manipulation which makes use of the Einstein relations for the diffusion coefficient for electrons and holes $qD_{(n,p)} = kT\mu_{(n,p)}$ and the fact that $dE_0/dx = 0$, the set of basic semiconductor equations in 1-D becomes (6.11 to 6.15)

$$\frac{d}{dx}\left(\varepsilon\frac{d\psi}{dx}\right) = -\rho \tag{6.11}$$

$$\frac{1}{q}\frac{dJ_n}{dx} + G_{opt} - R_{net} = 0 \tag{6.12}$$

$$-\frac{1}{q}\frac{dJ_p}{dx} + G_{opt} - R_{net} = 0 \tag{6.13}$$

$$J_n = qD_n\frac{dn}{dx} + \mu_n n\left[-q\frac{d\psi}{dx} - \frac{d\chi}{dx} - \frac{kT}{N_C}\frac{dN_C}{dx}\right] \tag{6.14}$$

$$J_p = -qD_p\frac{dp}{dx} + \mu_p p\left[-q\frac{d\psi}{dx} - \frac{d\chi}{dx} - \frac{dE_G}{dx} + \frac{kT}{N_V}\frac{dN_V}{dx}\right] \tag{6.15}$$

The space charge density ρ which appears in Poisson's equation (6.11) is in amorphous semiconductor material given by:

$$\rho = q\left(p - n + p_{loc} - n_{loc} + N_{don} - N_{acc}\right) \tag{6.16}$$

which takes into account free and localized charge carriers and ionized donors and acceptors, respectively. It should be mentioned, that not all incorporated dopant atoms in a-Si:H are electronically active at room temperature as is the case in crystalline silicon. Therefore N_{don} and N_{acc} are not the concentrations of dopant atoms incorporated in a-Si:H but only the ionized (active) ones. As it is difficult to measure directly the number of active dopants, commonly the measurement of the activation energy of the dark conductivity is used to approximate the position of the Fermi level in doped materials (see Section 3.1.1).

The primary function of a computer program is to solve this set of coupled partial differential equations. In order to solve the semiconductor equations one can choose between different sets of independent variables. Most commonly the electrostatic potential ψ and the free carrier concentrations p and n are used. Another set of variables is the electrical potential ψ and the quasi-Fermi levels for the electrons E_{FN} and holes E_{FP}. When solving these equations the proper values of the position dependent mobilities, space charge, recombination and generation rates and the boundary conditions must be known or evaluated.

In Eqns. 6.14 and 6.15 the first term represents the current due to diffusion of the carriers and the other term represents drift transport. The terms in the square brackets can be considered as the effective drift fields. In addition to the electric field they contain terms which reflect discontinuities or inhomogeneities in the material properties.

6.3.5 *Boundary conditions*

The solution of the set of coupled partial differential model equations (6.11 to 6.15) is mainly determined by the applied boundary conditions. There are two physical boundaries in 1-D device modeling, the front and the back contact of the device. The conditions at the contacts define the values of the model-independent variables at these points. The boundary conditions depend on the nature of the contacts of the device. Generally, two types of contacts can be distinguished: Ohmic contacts or Schottky contacts.

Ohmic contacts. In case of ideal Ohmic contacts, infinite surface recombination and charge neutrality at the contacts of the device (x = 0 and x = L, where L is the device thickness) is assumed

$$\rho\,(x=0) \quad = 0 \tag{6.17a}$$

$$\rho\,(x=L) \quad = 0 \tag{6.17b}$$

The space charge density $\rho(x)$ is given by equation (6.16). At the boundaries, the following conditions are introduced for the electrostatic potential ψ

$$\psi\,(x=0) \quad = \quad \psi_0\,(x=0) + V_{app} \tag{6.18a}$$

$$\psi\,(x=L) \quad = \quad \psi_0\,(x=L) \tag{6.18b}$$

where $\psi_0(x=0)$ and $\psi_0(x=L)$ are the solutions of equations 6.17a and 6.17b. respectively, and V_{app} is the applied external voltage.

In general, for Ohmic contacts the surface recombination velocities for electrons, S_n, and for holes, S_p, determine the carrier concentrations at the boundaries. The electron and hole current densities for this type of contacts are described in the following way

$$J_n\,(x=0) \quad = qS_{n0}\left[n\,(x=0) - n_{eq}\,(x=0)\right] \tag{6.19a}$$

$$J_n\,(x=L) \quad = qS_{nL}\left[n\,(x=L) - n_{eq}\,(x=L)\right] \tag{6.19b}$$

$$J_p\,(x=0) \quad = qS_{p0}\left[p\,(x=0) - p_{eq}\,(x=0)\right] \tag{6.20a}$$

$$J_p\,(x=L) \quad = qS_{pL}\left[p\,(x=L) - p_{eq}\,(x=L)\right] \tag{6.20b}$$

where n and p are the electron and hole concentrations at the contacts, n_{eq} and p_{eq} are the electron and hole concentrations in thermodynamic equilibrium at the front and back contacts. S_{n0}, S_{nL} are the surface recombination velocities for electrons at the front and back contact, respectively, and S_{p0}, S_{pL} are the surface recombination velocities for holes at the front and back contact, respectively.

Schottky contacts. In the case of Schottky contacts it is assumed that at thermal equilibrium the position of the Fermi level, E_F, at the contact depends on the effective barrier height, φ_b, of the metal semiconductor interface. This barrier is given by the difference between the work function of the metal, φ_m, and the electron affinity of the semiconductor, χ_s.

$$\varphi_b = \varphi_m - \chi_s = E_C - E_F \tag{6.21}$$

Once the position of the Fermi level at the interface as a function of φ_b has been determined, the concentration of electrons and holes at the interface is calculated from

$$n_{eq} = N_C \exp\left(-\frac{q\varphi_b}{kT}\right) \tag{6.22}$$

$$p_{eq} = n_i^2 / n_{eq} \tag{6.23}$$

where n_i is the intrinsic carrier concentration of the semiconductor material.

The current transport over the Schottky barrier is dominated by the majority carriers and is governed by the mechanism of thermionic emission (Rhoderick, 1980). The expression for the current density according to the thermionic-emission theory is

$$J = A^* T^2 \exp\left(-\frac{q\varphi_b}{kT}\right)\left[\exp\left(\frac{qV_{app}}{kT}\right) - 1\right] \tag{6.24}$$

where A^* is Richardson's constant. It can easily be shown that the current at the barrier boundary can be described by equations 6.19a and 6.19b with $S_n = A^* T^2 / qN_C$ in case of electrons and by equations 6.20a and 6.20b with $S_p = A^* T^2 / qN_V$ in case of holes.

6.4 DENSITY OF STATES MODELS FOR AMORPHOUS SILICON

6.4.1 Standard model of density of states distribution in a-Si:H

In order to simulate accurately the electrical and optical properties of a-Si:H devices, a proper description of the DOS distribution as a function of energy is of major importance. The present-day models of the electron states in the band gap of a-Si:H are based on the Cohen-Fritsche-Ovshinsky model (Cohen et al., 1969) and its modifications that include the dangling bond states (Mott and Davis, 1979, Street, 1991). In this section a general standard model of the DOS distribution in a-Si:H is introduced. The standard model of the DOS distribution consists of a parabolic conduction band (CB) and a parabolic valence band (VB), an exponentially decaying conduction band tail (CBT), an exponentially decaying valence band tail (VBT). Two equal distributions of

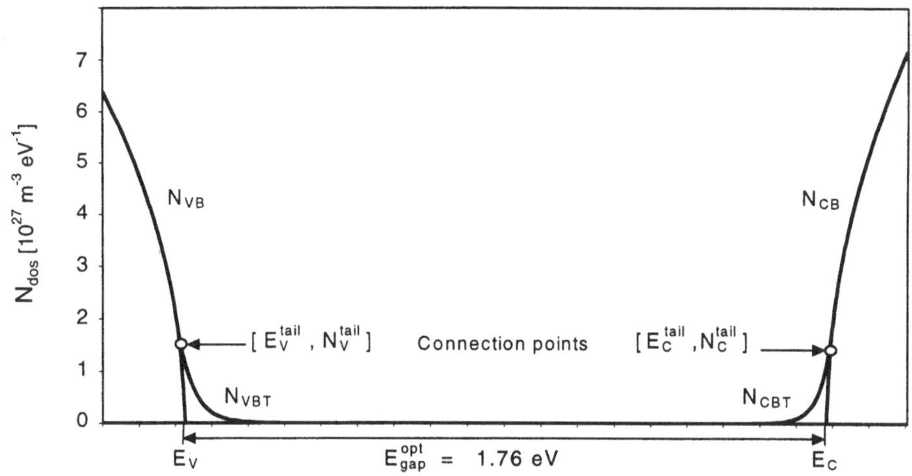

Figure 6.2. The standard model of the DOS distribution in a-Si:H on a linear scale.

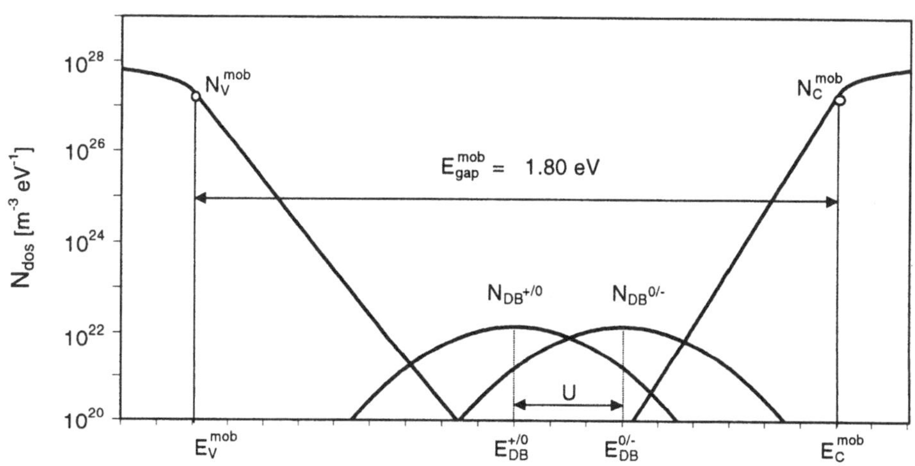

Figure 6.3. The standard model of the DOS distribution in a-Si:H on a logarithmic scale.

states around the midgap that are separated from each other by a correlation energy, represent the defect states related to dangling bonds ($DB^{+/0}$ and $DB^{0/-}$). The standard model of the DOS distribution in a-Si:H on a linear scale is shown in Fig. 6.2 and on a logarithmic scale in Fig. 6.3. The mathematical description of the DOS distribution in a-Si:H is given by the following formulas.

Conduction and valence band states. The DOS distribution of the conduction band

$$N_{\mathrm{CB+CBT}}(E) = N_{\mathrm{CB}}(E) \text{ for } E \geq E_{\mathrm{C}}^{tail} \tag{6.25a}$$
$$N_{\mathrm{CB+CBT}}(E) = N_{\mathrm{CBT}}(E) \text{ for } E \leq E_{\mathrm{C}}^{tail} \tag{6.25b}$$

$$N_{\mathrm{CB}}(E) = N_{\mathrm{C}}^{0}(E - E_{\mathrm{C}})^{1/2} \tag{6.26}$$

$$N_{\mathrm{CBT}}(E) = N_{\mathrm{C}}^{tail} \exp\left[-\left(\frac{E_{\mathrm{C}}^{tail} - E}{E_{\mathrm{C0}}^{tail}}\right)\right] \tag{6.27}$$

The DOS distribution of the valence band:

$$N_{\mathrm{VB+VBT}}(E) = N_{\mathrm{VB}}(E) \text{ for } E \leq E_{\mathrm{V}}^{tail} \tag{6.28a}$$
$$N_{\mathrm{VB+VBT}}(E) = N_{\mathrm{VBT}}(E) \text{ for } E \geq E_{\mathrm{V}}^{tail} \tag{6.28b}$$

$$N_{\mathrm{VB}}(E) = N_{\mathrm{V}}^{0}(E_{\mathrm{V}} - E)^{1/2} \tag{6.29}$$

$$N_{\mathrm{VBT}}(E) = N_{\mathrm{V}}^{tail} exp\left[-\left(\frac{E - E_{\mathrm{V}}^{tail}}{E_{\mathrm{V0}}^{tail}}\right)\right] \tag{6.30}$$

The energy levels E_{C} and E_{V} define the band gap. In the case of a-Si:H this band gap corresponds to the optical band gap E_{gap}^{opt}.

$$E_{gap}^{opt} = E_{\mathrm{C}} - E_{\mathrm{V}} \tag{6.31}$$

The parameter N_{C}^{0} (N_{V}^{0}) describes the parabolic distribution of states in the conduction (valence) band, E_{C0}^{tail} (E_{V0}^{tail}) the characteristic decay energy of the CB (VB) tail. ($E_{\mathrm{C}}^{tail}, N_{\mathrm{C}}^{tail}$) and ($E_{\mathrm{V}}^{tail}, N_{\mathrm{V}}^{tail}$) are the connection points of the parabolic and the exponential part of the conduction and the valence band, respectively.

Dangling bond states. In order to reflect the experimentally observed continuous distribution of dangling bond (DB) states in the band gap, a Gaussian distribution is used in the standard model to describe the DB states distribution. A dangling bond can be in three charge states: positive (D^{+}), neutral (D^{0}) and negative (D^{-}). An imperfection with three possible charge states acts to a good approximation like a group of two imperfections consisting of a donor-like state ($\mathrm{DB}^{+/0}$) and an acceptor-like state ($\mathrm{DB}^{0/-}$) and is therefore represented by two energy levels $E^{+/0}$ and $E^{0/-}$ in the band diagram, respectively. These energy levels are called the transition energy levels. The transition energy levels are separated from each other by a correlation energy,

U, which is the energy needed to add the second electron to a singly occupied (neutral) dangling bond. The correlation energy is assumed to be constant and positive. Under these assumptions the dangling bonds are represented by two equal Gaussian distributions in the band diagram separated from each other by the correlation energy. The mathematical description of the density of DB states is given by

$$N_{\mathrm{DB}^{+/0}}(E) = \frac{N_{\mathrm{DB}}^{tot}}{\sigma_{db}\sqrt{2\pi}} \exp\left[-\frac{\left(E-E_{\mathrm{DB}}^{+/0}\right)^2}{\left(2\sigma_{db}^2\right)}\right] \tag{6.32a}$$

$$N_{\mathrm{DB}^{0/-}}(E) = \frac{N_{\mathrm{DB}}^{tot}}{\sigma_{db}\sqrt{2\pi}} \exp\left[-\frac{\left(E-E_{\mathrm{DB}}^{0/-}\right)^2}{\left(2\sigma_{db}^2\right)}\right] \tag{6.32b}$$

$$N_{\mathrm{DB}^{0/-}}(E) = N_{\mathrm{DB}^{+/0}}(E+U) \tag{6.33}$$

$$E_{\mathrm{DB}}^{0/-} = E_{\mathrm{DB}}^{+/0} + U \tag{6.34}$$

N_{DB}^{tot} is the total density of dangling bonds, $E_{\mathrm{DB}}^{+/0}$ and $E_{\mathrm{DB}}^{0/-}$ is the energy of the peak of the Gaussian distribution for the donor-like states $\mathrm{DB}^{+/0}$ and the acceptor-like states $\mathrm{DB}^{0/-}$, respectively, σ_{db} is the standard deviation of the distribution, and U is the correlation energy.

For modeling purposes the continuous distribution of states in the band gap is mapped on a set of discrete energy levels E_m with a corresponding number of energy states N_m. The number of energy states in the energy interval $(E_m - \ E/2, E_m + \ E/2)$ can be calculated by

$$N_m = \int_{E_m-\frac{\Delta E}{2}}^{E_m+\frac{\Delta E}{2}} N(E)dE \tag{6.35}$$

where $N(E)$ is the density of states and $\Delta E = E_m - E_{m-1}$.

6.4.2 *Extended and localized states in a-Si:H*

When considering the transport properties of carriers in a-Si:H we have to distinguish between the extended states and the localized states in the DOS distribution. The energy levels E_{C}^{mob} and E_{V}^{mob}, which are called the mobility edge of the conduction band and the mobility edge of the valence band, respectively, define a mobility gap E_{gap}^{mob} (see Fig. 6.3).

$$E_{gap}^{mob} = E_{\mathrm{C}}^{mob} - E_{\mathrm{V}}^{mob} \tag{6.36}$$

The localized states within the mobility gap consist of the tail states and the defect states. These states are different in nature. The tail states behave like ordinary acceptor-like states (CB tail states) or donor-like states (VB tail states). The dangling bonds are amphoteric, i.e., they can act both as acceptor-like and donor-like states and are represented by two energy levels. The different nature of the two types of localized states requires that different models are applied to calculate the R-G rate through these states. It is assumed that within the mobility gap the mobility of charge carriers is zero.

The states in the conduction band above the mobility edge E_{C}^{mob} are extended. They are populated with electrons that are characterized by a concentration n and an extended-state mobility $\mu_n > 0$. The extended states in the valence band below the mobility edge E_{V}^{mob} are populated with holes that are characterized by a concentration p and an extended-state mobility $\mu_p > 0$. It should be noted that the mobility edges need not be identical to the band edges (E_{C}, E_{V}) or to the connection points (E_{C}^{tail}, E_{V}^{tail}), though most of the modeling groups define the mobility edges equal to the connection points $E_{\mathrm{C}}^{mob} = E_{\mathrm{C}}^{tail}$, $E_{\mathrm{V}}^{mob} = E_{\mathrm{V}}^{tail}$. The values of the DOS at the mobility edges E_{C}^{mob} and E_{V}^{mob} are denoted N_{C}^{mob} and N_{V}^{mob}, respectively.

In the operational range of a-Si:H solar cells the dominant transport mechanism is multiple trapping. The transport is then characterized by the concentration of carriers in the extended states (n, p) and the extended-state mobilities (μ_n, μ_p). Further it is assumed that the mobilities are the same in thermal equilibrium and in steady state.

For the concentration of carriers in the extended states one can write

$$n = N_{\mathrm{C}mob}^{eff} \exp\left(\frac{E_{\mathrm{FN}} - E_{\mathrm{C}}^{mob}}{kT}\right) \tag{6.37a}$$

$$p = N_{\mathrm{V}mob}^{eff} \exp\left(\frac{E_{\mathrm{V}}^{mob} - E_{\mathrm{FP}}}{kT}\right) \tag{6.37b}$$

where $N_{\mathrm{C}mob}^{eff}$ ($N_{\mathrm{V}mob}^{eff}$) is the effective density of states at CB (VB) mobility edge.

6.4.3 Defect pool model for dangling bond states distribution

The Gaussian distribution that is used to describe the DB states distribution is based on the idea that the structural disorder leads to a distribution of energy states in the band gap. It does not contain any information about the origin of these states. The defect pool theory, which began to take shape in the beginning of the nineties (Winer, 1990, Powell and Deane, 1993), has attracted a lot of attention since it could successfully describe the defect structure in both doped and undoped a-Si:H in thermal equilibrium as well as the metastable defect formation in non-equilibrium (Schumm, 1994).

This theory is based on the weak-bond - dangling-bond conversion model (Stutzmann, 1987). In thermodynamical equilibrium chemical-type reactions take place between the intrinsic defects (the dangling bonds) and defect-free coordinations (the Si-Si weak bonds and/or Si-H bonds). The weak Si-Si bonds are associated with the states in the VB tail. Hydrogen atoms are involved in the defect creation reaction and mediate the separation of two dangling bonds that are created when a weak Si-Si bond is broken. The mean energy of the electrons in the DB states depends on the charge state of the dangling bond. Since the charge state depends on the position of the Fermi level, the Fermi level influences the dangling bond distribution. The equilibrium DB states distribution is calculated by minimizing the entropy of the system that is formed by silicon weak bonds, dangling bonds and silicon-hydrogen bonds. The resulting DB distribution is known as the defect-pool model (DPM).

Most of the modeling groups use the defect-pool model of the DB states distribution as published by Powell and Deane, 1993. Their derivation of the DB states distribution is based on the following assumptions:

1. The energy distribution of the weak Si-Si bonds corresponds to the VB tail states distribution (Eqn. 6.30).

2. The distribution of the available defect sites P(E), which is called the defect pool, is described by a Gaussian distribution

$$P(E) = \frac{1}{\sigma_{dp}\sqrt{2}} \exp\left[-\frac{(E - E_{dp})^2}{2\sigma_{dp}^2}\right] \tag{6.38}$$

where σ_{dp} is the standard deviation and E_{dp} is the peak position of the Gaussian distribution or the so-called defect pool center. The resulting expression for the DB states distribution in equilibrium is

$$N_{\mathrm{DB}^{+/0}}(E) = \gamma \left[\frac{2}{F_{eq}^{0}(E)}\right]^{\rho kT / E_{\mathrm{V0}}^{tail}} P\left(E + \frac{\rho\sigma_{pd}^2}{E_{\mathrm{V0}}^{tail}}\right) \tag{6.39}$$

where F_{eq}^{0} is the equilibrium occupation function of neutral dangling bond state, γ is the scaling factor. The thermal equilibrium occupation functions of the dangling bond are (Okamoto et al., 1984)

$$F_{eq}^{+} = \frac{1}{Z} \tag{6.40a}$$

$$F_{eq}^{0} = \frac{2}{Z}\exp\left(\frac{E_{\mathrm{F}} - E}{kT}\right) \tag{6.40b}$$

$$F_{eq}^{-} = \frac{1}{Z}\exp\left(\frac{2E_{\mathrm{F}} - 2E - U}{kT}\right) \tag{6.40c}$$

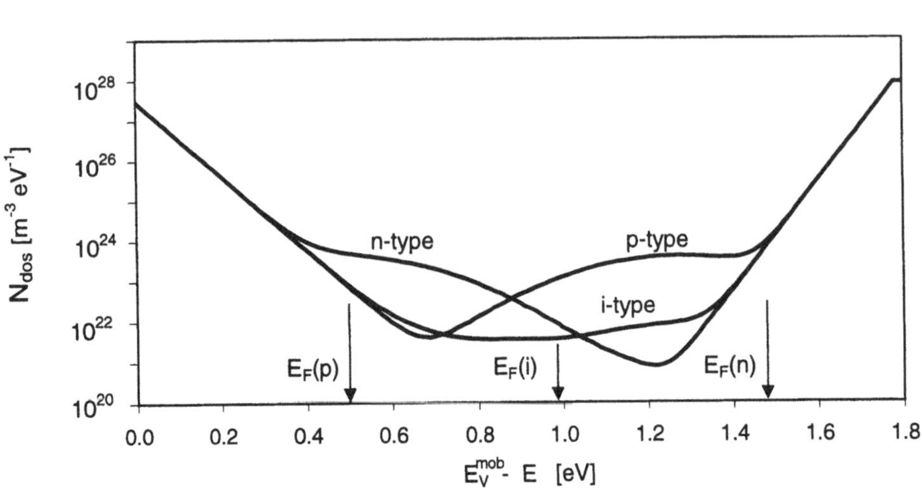

Figure 6.4. The DOS distribution in the mobility gap of intrinsic a-Si:H at different positions in the solar cell according to the defect pool model for the single electron DB states distribution. Three cases with different position of Fermi level are illustrated: near the p-type layer (p-type and $E_{\rm F}$(p)), in the center of the intrinsic layer (i-type and $E_{\rm F}$(i)), and near the n-type layer (n-type and $E_{\rm F}$(n)).

$$Z = 1 + 2\exp\left(\frac{E_{\rm F} - E}{kT}\right) + \exp\left(\frac{2E_{\rm F} - 2E - U}{kT}\right) \tag{6.40d}$$

For γ the following expression is derived

$$\gamma = \left[\frac{N_{\rm V}^{tail}2\left(E_{\rm V0}^{tail}\right)^2}{\left(2E_{\rm V0}^{tail} - kT\right)}\right]^{\rho}\left[\frac{i}{2c_{\rm H}}\right]^{\rho-1}\exp\left(\frac{-\rho}{E_{\rm V0}^{tail}}\left[E_{dp} - E_{\rm V}^{tail} - \frac{\rho\sigma_{pd}^2}{2E_{\rm V0}^{tail}}\right]\right) \tag{6.41}$$

where $c_{\rm H}$ is the total hydrogen concentration in a-Si:H and i indicates the number H atoms mediating the weak-bond-breaking chemical reaction (i=0, 1 or 2). The parameter ρ depends on the number i in the following way.

$$\rho = \frac{2E_{\rm V0}^{tail}}{\left(2E_{\rm V0}^{tail} + ikT\right)} \tag{6.42}$$

Since the DB distribution depends on the position of the Fermi level in the device, the DB distribution profile in the solar cell is calculated self-consistently at the freeze-in temperature (usually the deposition temperature) by an iterative method. In this procedure the set of model equations (6.11-6.15) is solved together with the expression for the dangling bond distribution (6.39). The freeze-in temperature is higher than the operational temperature of the device. The temperature dependence of the characteristic decay energy of the VB tail is taken into account by using the following relationship (Stutzmann, 1992).

$$E_{\mathrm{V0}}^{tail^2}(T) = E_{\mathrm{V0}}^{tail^2}(T=0) + (kT)^2 \tag{6.43}$$

Fig. 6.4 shows the DOS distribution in the mobility gap of the intrinsic a-Si:H at different positions in the solar cell. The DB states distribution is calculated with the defect pool model of Powell and Deane, 1993. The figure demonstrates the strong influence of the position of Fermi level on the distribution of the DB states. The position of the Fermi level determines the total number of defects and the peak position of the DB distribution. The theory predicts that the total number of dangling bonds in the intrinsic a-Si:H increases when the Fermi level shifts from mid-gap towards the mobility edges of the bands.

6.5 RECOMBINATION-GENERATION STATISTICS IN AMORPHOUS SILICON

6.5.1 R-G statistics of the continuously distributed states in the band gap

In crystalline semiconductors the recombination process is typically dominated by recombination centers at a single energy level in the band gap. In a-Si:H, the band gap is characterized by a continuous density of allowed states that in its entirety contributes to the net R-G rate. Hence, it is necessary to add up or integrate the recombination rate contributions from the gap states over the energy band gap. Under the assumption that the recombination centers are non-interacting the net R-G rate can be calculated from

$$R_{net} = \int_{E_{\mathrm{V}}^{mob}}^{E_{\mathrm{C}}^{mob}} N(E)\, \eta_{\mathrm{R}}(E)\, dE \tag{6.44}$$

where $\eta_R(E)$ is the recombination rate contribution of a state at energy E, and $N(E)$ is the density of states as a function of energy in the band gap. For the calculation of the trapped charge in the band gap states we have to distinguish between the tail states and the dangling bond states. The trapped charge in the tail states is calculated using the following equations for ordinary donor-like and acceptor-like states, respectively,

$$\rho_{\mathrm{D}} = q \int_{E_{\mathrm{C}}^{mob}}^{E_{\mathrm{V}}^{mob}} N_{\mathrm{D}}(E)\,[1 - f(E)]\, dE \tag{6.45}$$

$$\rho_{\mathrm{A}} = -q \int_{E_{\mathrm{C}}^{mob}}^{E_{\mathrm{V}}^{mob}} N_{\mathrm{A}}(E) f(E) \, dE \tag{6.46}$$

where $f(E)$ is the occupation function. In case of the amphoteric DB states the space charge is given by

$$\rho_{\mathrm{DB}} = q \int_{E_{\mathrm{C}}^{mob}}^{E_{\mathrm{V}}^{mob}} N_{\mathrm{DB}}(E) \left[F^{+}(E) - F^{-}(E)\right] \, dE \tag{6.47}$$

In this equation $F^{+}(E)$ and $F^{-}(E)$ are the occupation functions of empty and doubly occupied dangling bonds, respectively.

The assumption of non-interacting centers means that the localized states in the band gap can interact only with carriers in the extended states of the conduction and valence bands. This assumption allows us to use the Shockley-Read-Hall (SRH) R-G statistics (Shockley and Read, 1952) to model the recombination process through the single-level states and the multilevel R-G statistics (Sah and Shockley, 1958) for the amphoteric DB states. Fig. 6.5 schematically depicts the different types of the localized states in the band gap of a-Si:H and the models that are used to calculate the recombination rate through these states and their charge occupation. The full SRH statistics can be simplified by the approach of Simmons and Taylor, 1971. The full treatment of R-G statistics of the amphoteric DB states using the Sah and Shockley, 1958 approach is often replaced by the decoupled model of dangling bonds on which the Taylor and Simmons approach of R-G statistics is applied. In the decoupled model a dangling bond is represented by a pair of uncorrelated donor-like and acceptor-like states (Halpern, 1986). A detailed analysis and comparison of these approaches is given by Willemen, 1998.

6.5.2 R-G statistics of VB and CB tail states

The VB tail states behave like donor states and are neutral (T_{D}^{0}) or positive (T_{D}^{+}). The CB tail states behave like acceptor states and are neutral (T_{A}^{0}) or negative (T_{A}^{-}). In the energy band diagram the donor and acceptor R-G center is represented by a single energy level. The theory that describes the recombination process involving a single energy level in the band gap of semiconductor was developed by Shockley and Read, 1952 and Hall, 1951, Hall, 1952.

This theory is based on four possible transitions between an energy level E_{T} in the band gap and the extended states of the conduction and valence bands:

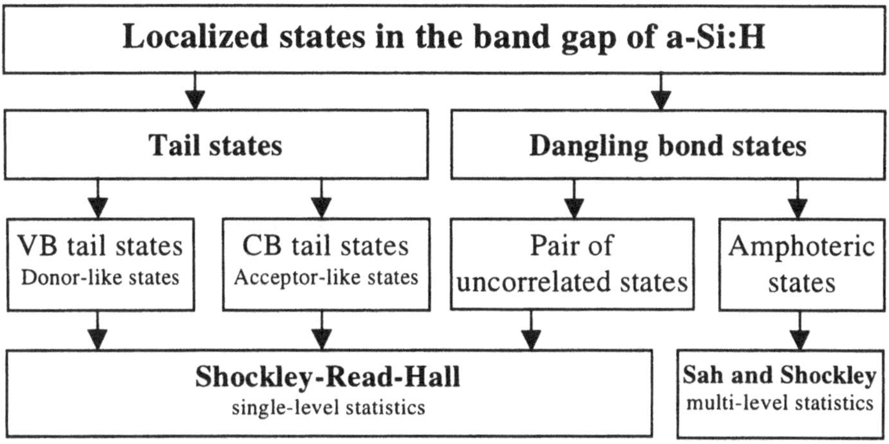

Figure 6.5. Different types of the localized states in the band gap of a-Si:H and the models that are used to calculate the recombination rate and charge occupation (after Willemen, 1998).

1) electron capture at an R-G center
2) electron thermal emission from an R-G center
3) hole capture at an R-G center
4) hole thermal emission from an R-G center

These processes are schematically shown in Fig. 6.6 and their corresponding rates of change in the carrier concentrations are listed in Table 6.1. A general derivation of the steady-state recombination rate and occupation function involving single-level states at energy E_{T}, which result from the processes in Table 6.1 is given in Pierret, 1987.

VB tail states. The resulting expression for the recombination rate of the VB tail states at energy E_{T} is

$$R_{\mathrm{VB}} = N_{\mathrm{D}} \frac{C_p^0 C_n^+ \left(np - n_i^2\right)}{C_p^0 n + e_p^+ + C_n^+ p + e_n^0} \tag{6.48}$$

In this equation, N_{D} is the total number of donor-like states at energy E_{T}, C_p^0 (C_n^+) is the capture-rate coefficient for holes (electrons), and e_p^+ (e_n^0) is the emission coefficient for holes (electrons). The emission coefficients for electrons and holes are determined from the principle of detailed balance (Pierret, 1987). This principle states that under equilibrium conditions each fundamental process and its inverse must self-balance independent of any other process that

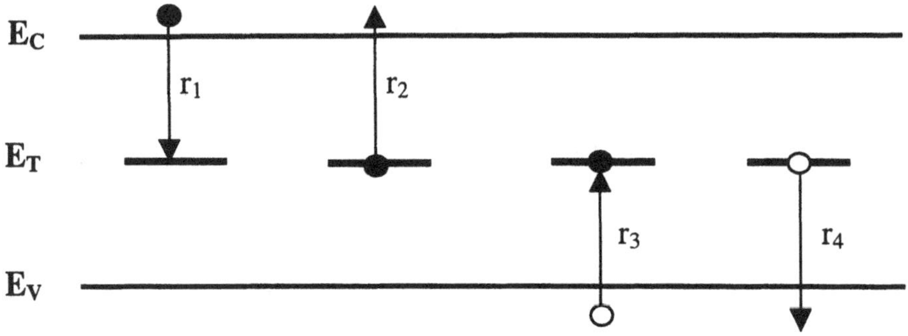

Figure 6.6. Electronic transitions in the recombination process between a single energy level in the band gap of semiconductor and the energy bands.

may be occurring inside the material. Assuming that the emission and capture coefficients under nonequilibrium conditions remain approximately equal to their equilibrium values, the emission coefficients for electrons and holes can be expressed as

$$e_n^0 = C_n^+ N_C \exp\left[(E_T - E_C)/kT\right] \quad (6.49a)$$
$$e_p^+ = C_p^0 N_V \exp\left[(E_V - E_T)/kT\right] \quad (6.49b)$$

Table 6.1. The time rates of change in the carrier concentrations for transitions involving VB and CB tail states at arbitrary energy level E_T in the band gap.

	VB tail states (donor-like R-G centers)		*CB tail states (acceptor-like R-G centers)*	
	Transition	*Rate*	*Transition*	*Rate*
r_1	$T_D^+ + e \rightarrow T_D^0$	$n \cdot C_n^+ \cdot N_T \cdot (1 - f^0)$	$T_A^0 + e \rightarrow T_A^-$	$n \cdot C_n^0 \cdot N_T \cdot (1 - f^-)$
r_2	$T_D^0 \rightarrow T_D^+ + e$	$e_n^0 \cdot N_T \cdot f^0$	$T_A^- \rightarrow T_A^0 + e$	$e_n^- \cdot N_T \cdot f^-$
r_3	$T_D^0 + h \rightarrow T_D^+$	$p \cdot C_p^0 \cdot N_T \cdot f^0$	$T_A^- + h \rightarrow T_A^0$	$p \cdot C_p^- \cdot N_T \cdot f^-$
r_4	$T_D^+ \rightarrow T_D^0 + h$	$e_p^+ \cdot N_T \cdot (1 - f^0)$	$T_A^0 \rightarrow T_A^- + h$	$e_p^0 \cdot N_T \cdot (1 - f^-)$

r_1 is electron capture, r_2 is electron emission, r_3 is hole capture, and r_4 is hole emission.

In the case of a-Si:H the effective density of states N_{C} and N_{V} are equal to $N_{\mathrm{C}mob}^{eff}$ and $N_{\mathrm{V}mob}^{eff}$, respectively. In thermal equilibrium the occupation function of states at energy E_{T} is described by the Fermi-Dirac distribution function. The steady-state occupation function of donor-like states is given by

$$f^0 = \frac{C_n^+ n + e_p^+}{C_n^+ n + e_n^0 + C_p^0 p + e_p^+} \tag{6.50}$$

Since the density of VB tail states continuously decays into the band gap the total R-G rate and the concentration of localized holes including all VB tail states is calculated from Eqns. 6.44 and 6.45, respectively. The integrals in these equations are replaced by a summation over a set of m discrete energy levels on which the continuous density of VB tail states is mapped

$$R_{\mathrm{VB}}^{tot} = \sum_m R_{\mathrm{VB}} \tag{6.51}$$

$$p_{\mathrm{VB}}^{tot} = \sum_m N_{\mathrm{D}} \left(1 - f^0\right) \tag{6.52}$$

CB tail states. The derivation of the recombination rate and steady state occupation function of the CB tail states is analogous to the VB tail states. The resulting formulas are:

$$R_{\mathrm{CB}} = N_{\mathrm{A}} \frac{C_p^- C_n^0 \left(np - n_i^2\right)}{C_p^- n + e_p^0 + C_n^0 p + e_n^-} \tag{6.53}$$

$$e_n^- = C_n^0 N_{\mathrm{C}} \exp\left[(E_{\mathrm{T}} - E_{\mathrm{C}})/kT\right] \tag{6.54a}$$

$$e_p^0 = C_p^- N_{\mathrm{V}} \exp\left[(E_{\mathrm{V}} - E_{\mathrm{T}})/kT\right] \tag{6.54b}$$

$$f^- = \frac{C_n^0 n + e_p^0}{C_n^0 n + e_n^- + C_p^- p + e_p^0} \tag{6.55}$$

where N_{A} is the total number of acceptor-like states at energy E_{T}, C_p^- (C_n^0) is the capture-rate coefficients for holes (electrons), and e_p^0 (e_n^-) is the emission coefficients for holes (electrons). The total R-G rate and concentration of localized electrons including all CB tail states is calculated from Eqs. 6.44 and 6.46, respectively, replacing the integrals by a summation:

$$R_{\mathrm{CB}}^{tot} = \sum_m R_{\mathrm{CB}} \tag{6.56}$$

$$n_{\mathrm{CB}}^{tot} = \sum_m N_{\mathrm{T}} f^- \tag{6.57}$$

6.5.3 R-G statistics of DB states

A dangling bond can be in three charge states: positive (D^+), neutral (D^0) and negative (D^-). In the band diagram a dangling bond is represented by two energy levels $E^{+/0}$ and $E^{0/-}$. The R-G statistics involving the two-level representation of dangling bonds is correctly described by the Sah and Shockley multilevel R-G statistics. Possible electronic transitions between an amphoteric R-G center and the energy bands are illustrated in Fig. 6.7 and the rates of change in carrier concentrations due to these transitions are listed in Table 6.2.

The probability that a dangling bond is in a positive, neutral or negative charge state is given by the occupation functions F^+, F^0, and F^-, respectively. The sum of the occupation functions has to be unity.

$$F^+ + F^0 + F^- = 1 \tag{6.58}$$

The emission coefficients are determined from the principle of detailed balance applied to the processes listed in Table 6.2 and the thermal equilibrium occupation functions (Eqns. 6.40a-6.40d).

$$e_n^0 = C_n^+ N_C \exp\left[\frac{(E - kT\ln 2) - E_C}{kT}\right] \tag{6.59a}$$

$$e_p^+ = C_p^0 N_V \exp\left[\frac{E_V - (E - kT\ln 2)}{kT}\right] \tag{6.59b}$$

$$e_n^- = C_n^0 N_C \exp\left[\frac{(E + U + kT\ln 2) - E_C}{kT}\right] \tag{6.59c}$$

$$e_p^0 = C_p^- N_V \exp\left[\frac{E_V - (E + U + kT\ln 2)}{kT}\right] \tag{6.59d}$$

Under steady state conditions the charge occupation of the DB states does not

Table 6.2. The time rates of change in the carrier concentrations for recombination processes involving DB states.

	$DB^{+/0}$ states Transition	Rate		*$DB^{0/-}$ states* Transition	Rate
$\mathbf{r}_1$	$D^+ + e \to D^0$	$n \cdot C_n^+ \cdot N_{DB} \cdot F^+$	$\mathbf{r}_5$	$D^0 + e \to D^-$	$n \cdot C_n^0 \cdot N_{DB} \cdot F^0$
$\mathbf{r}_2$	$D^0 \to D^+ + e$	$e_n^0 \cdot N_{DB} \cdot F^0$	$\mathbf{r}_6$	$D^- \to D^0 + e$	$e_n^- \cdot N_{DB} \cdot F^-$
$\mathbf{r}_3$	$D^0 + h \to D^+$	$p \cdot C_p^0 \cdot N_{DB} \cdot F^0$	$\mathbf{r}_7$	$D^- + h \to D^0$	$p \cdot C_p^- \cdot N_{DB} \cdot F^-$
$\mathbf{r}_4$	$D^+ \to D^0 + h$	$e_p^+ \cdot N_{DB} \cdot F^+$	$\mathbf{r}_8$	$D^0 \to D^- + h$	$e_p^0 \cdot N_{DB} \cdot F^0$

$\mathbf{r}_1$, $\mathbf{r}_5$ is electron capture, $\mathbf{r}_2$, $\mathbf{r}_6$ is electron emission, $\mathbf{r}_3$, $\mathbf{r}_7$ is hole capture, $\mathbf{r}_4$, $\mathbf{r}_8$ is hole emission.

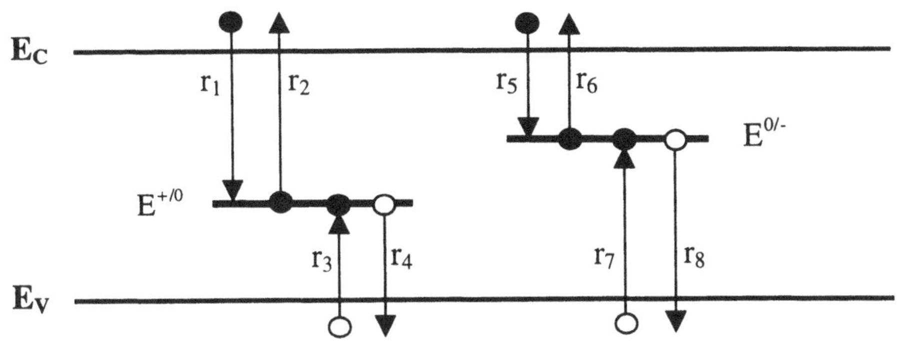

Figure 6.7. Possible electronic transitions in the recombination process between the energy bands and an amphoteric R-G center represented by two energy levels in the band gap.

change with time, which leads to the following equations

$$0 = -C_n^+ nF^+ N_{\mathrm{DB}} + e_n^0 F^0 N_{\mathrm{DB}} + C_p^0 pF^0 N_{\mathrm{DB}} - e_p^+ F^+ N_{\mathrm{DB}} \quad (6.60a)$$

$$0 = -C_n^0 nF^0 N_{\mathrm{DB}} + e_n^- F^- N_{\mathrm{DB}} + C_p^- pF^- N_{\mathrm{DB}} - e_p^0 F^0 N_{\mathrm{DB}} \quad (6.60b)$$

where N_{DB} is the concentration of DB states with transition energies $E^{+/0}$ and $E^{0/-}$ transition energies.

After a bit of manipulation one can rewrite the Eqs. 6.60a and 6.60b

$$F^+ N^+ = F^0 P^0 \quad (6.61a)$$

$$F^0 N^0 = F^- P^- \quad (6.61b)$$

where

$$N^+ = C_n^+ n + e_p^+ \quad (6.62a)$$

$$P^0 = C_p^0 p + e_n^0 \quad (6.62b)$$

$$N^0 = C_n^0 n + e_p^0 \quad (6.62c)$$

$$P^- = C_p^- p + e_n^- \quad (6.62d)$$

Taking into account that the sum of the occupation functions is unity, the steady-state occupation functions can be described as follows

$$F^+ = \frac{P^- P^0}{N^+ P^- + P^0 P^- + N^+ N^0} \quad (6.63a)$$

$$F^0 = \frac{P^- N^+}{N^+P^- + P^0P^- + N^+N^0} \tag{6.63b}$$

$$F^- = \frac{N^0 N^+}{N^+P^- + P^0P^- + N^+N^0} \tag{6.63c}$$

The R-G rate involving the dangling bond states represented by a pair of energy levels $E^{+/0}$ and $E^{0/-}$ in the band gap is

$$R_{\mathrm{DB}} = N_{\mathrm{DB}} \left(np - n_i^2\right) \frac{C_n^+ C_p^0 P^- + C_n^0 C_p^- N^+}{N^+P^- + P^0P^- + N^+N^0} \tag{6.64}$$

The net contribution of all DB states to the R-G is calculated by using Eqn. 6.44. The integral in this equation is replaced by summation over a set of m discrete energy levels on which the continuous density of DB states is mapped

$$R_{\mathrm{DB}}^{tot} = \sum_m R_{\mathrm{DB}} \tag{6.65}$$

The total concentration of localized holes and electrons in DB states is calculated using the following expressions

$$p_{\mathrm{DB}}^{tot} = \sum_m N_{\mathrm{DB}} F^+ \tag{6.66a}$$

$$n_{\mathrm{DB}}^{tot} = \sum_m N_{\mathrm{DB}} F^- \tag{6.66b}$$

6.5.4 Total R-G rate and space charge in the localized states of the band gap

The net R-G rate including all states in the band gap is calculated as a sum of the net R-G rates of VB and CB tail states and DB states.

$$R_{net} = R_{\mathrm{VB}}^{tot} + R_{\mathrm{CB}}^{tot} + R_{\mathrm{DB}}^{tot} \tag{6.67}$$

The total space charge in the localized states in the band gap is

$$\rho_{loc}^{tot} = q\left(p_{\mathrm{VB}}^{tot} - n_{\mathrm{CB}}^{tot} + p_{\mathrm{DB}}^{tot} - n_{\mathrm{DB}}^{tot}\right). \tag{6.68}$$

6.6 NUMERICAL SOLUTION OF THE SEMICONDUCTOR EQUATIONS

Before the set of coupled partial differential model equations is numerically solved the device structure is divided into a number of small cells. This step is called a grid or mesh generation. The set of equations is then discretized on the generated grid points using the finite differences method and a linear interpolation method (Selberherr, 1984). The non-linear terms in the equations, such as the term containing the second derivative of the Poisson equation, are approximated by Taylor expansion series disregarding the higher order components. The spatial discretization of the set of the three partial differential equations over N grid points results in a set of $3 \times N$ nonlinear algebraic equations with $3 \times N$ unknowns. Generally, iterative methods are used to solve systems of nonlinear algebraic equations. One of the most often used method is Newton's method (Selberherr, 1984).

In order to solve the resulting set of $3 \times N$ algebraic equations two different approaches are used
(1) Gummel's method (often referenced as decoupled method) that solves 3 partly coupled sets of N equations using the iterative Newton method (Gummel, 1964, Selberherr, 1984). (2) the fully coupled method that solves simultaneously the set of $3 \times N$ equations applying Newton's iterative method (Selberherr, 1984, Kurata, 1982).

The advantages of Gummel's method are the relatively simple program code and the broad range of convergence, which allows the use of relatively poor initial guesses for the independent variables. However, the convergence properties become very poor, when the coupling between the equations is too strong. This is, for instance, the case with large generation-recombination terms, which occur frequently in a-Si:H solar cells. On the other hand, the fully coupled method shows a much better convergence rate in cases with strong recombination, where Gummel's method fails. However, the initial guess should be relatively close to the solution. In terms of program complexity and matrix computation time this algorithm is more elaborate.

References

Block, M., *Modellierung von Dünnschichtsolarzellen aus amorphem Silizium*, Ph.D. thesis, Fachbereich Physik der Phillips-Universität Marburg, 1993.

Bruns, J., *Die Entwicklung eines numerischen Simulationmodells für a-Si:H Solarzellen und seine Anwendung zur Analyse experimentell ermittelter Spektralcharakteristiken*, Ph.D. thesis, Fachbereich Elektrotechnik der Technischen Universität Berlin, 1993.

Chen, I., and S. Lee, *On the current-voltage characteristics of amorphous hydrogenated silicon Schottky diodes*, J. Appl. Phys. **53** (1982) 1045-1051.

Cohen, M.H., H. Fritsche, and S.R. Ovshinsky, *Simple band model for amorphous semiconducting alloys* , Phys. Rev. Lett., **22** (1969) 1065-1068.

Crandall, R.S., *Modelling of thin film solar cells: Uniform field approximation*, J. Appl. Phys. **54** (1983) 7176-7186.

Fantoni, A., M. Vieira, J. Cruz, R. Schwarz and R. Martins, *A two-dimensional numerical simulation of a non-uniformly illuminated amorphous silicon solar cell*, J. Physics D: Appl. Phys. **29** (1996) 3154-3159.

Fonash, S., and H. Zhu, *Computer simulation for solar cell applications: Understanding and design*, in: Amorphous and Microcrystalline Silicon Technology - 1998, edited by R. Schropp, H. Branz, S. Wagner, M. Hack, and I. Shimizu, Materials Research Society Symp. Proc. **507** (1998), in print.

Gummel, H.K., *A self-consistent iterative scheme for one-dimensional steady state transistor calculations*, IEEE Trans. ED, **ED-11** (1964) 455-465.

Hack, M. and M. Shur, *Physics of amorphous silicon alloy p-i-n solar cells*, J. Appl. Phys. **58** (1985) 997-1020.

Hall, R.N., *Germanium rectifier characteristics*, Phys. Rev. **83** (1951) 228.

Hall, R.N., *Electron-hole recombination in germanium*, Phys. Rev. **87** (1952) 387.

Halpern, V., *The statistics of recombination via dangling bonds in amorphous silicon*, Phil. Mag. **54** (1986) 473-482.

Hou, J.Y., J.K. Arch, S.J. Fonash, S. Wiedeman, and M. Bennet, *An examination of. the "tunneljunctions" in triple junction a-Si:H based solar cells: Modeling and effects on performance*, Proc. 22nd IEEE PV Specialists Conference, Las Vegas (1991) 1260-1264.

Hurkx, G.A.M., D.B.M. Klaassen and M.P.G. Knuvers, *A new recombination model for device simulation including tunneling*, IEEE Trans. ED **39** (1992) 331-338.

Ikegaki, T., H. Itoh, S. Muramatsu, S. Matsubara, N. Nakamura, T. Shimada, J. Umeda, and M. Migitaka, *Numerical analysis of amorphous silicon solar cells: A detailed investigation of the effects of internal field distribution on cell characteristics*, J. Appl. Phys. **58** (1985) 2352-2359.

Kroon, M.A., R.A.C.M.M. van Swaaij, M. Zeman, V.I. Kuznetsov, and J.W. Metselaar, *Hydrogenated amorphous silicon transverse junction solar cell*, Appl. Phys. Lett. **72** (1998) 209.

Kurata, M., *Numerical analysis for semiconductor devices,* (Lexington Books, Lexington, MA, 1982).

McElheny, P.J., J. Arch, H. Liu and S.J. Fonash, *Range of validity of the surface-photovoltage diffusion length measurement: A computer simulation*, J. Appl.Phys. **64** (3), 1254-1265 (1988).

MEDICI, Two-Dimensional Device Simulation Program, Version 4.0, TMA, Sunnyvale, CA, 1997.

Mott, N.F., and E.A. Davis, *Electronic processes in non-crystalline materials*, 2nd Edition (The International Series of Monographs on Physics, ed. W. Marshall and D.H. Wilkinson, Clarendon Press, Oxford, 1979).

Okamoto, H., H. Kida, S. Nonomura and Y. Hamakawa, *Variable minority carrier transport model for amorphous silicon solar cells*, Solar Cells **8** (1983) 317-336.

Okamoto, H., H. Kida and Y. Hamakawa, *Steady-state photoconductivity in amorphous semiconductors containing correlated defects*, Phil. Mag. **49** (1984) 231-247.

Overhof, H., and P. Thomas, *Hydrogenated amorphous silicon* (Springer-Verlag, 1989).

Pierret, R.F., in: Advanced Semiconductor Fundamentals (Addison-Wesley Publishing Co., Reading, Massachusetts, 1987).

Powell M.J., and S.C. Deane, *Improved defect-pool model for charged defects in amorphous silicon*, Phys. Rev. B **48** (1993) 10815-10827.

Rhoderick, E.H., *Metal-Semiconductor Contacts*, Monographs in Electrical and Electronic Engineering, Eds. P. Hammond and D.Walsh (Clarendon Press, Oxford, 1980).

Sah, C.T., and W. Shockley, *Electron-Hole recombination statistics in semiconductors through flaws with many charge conditions*, Phys. Rev. **109** (1958) 1103-1115.

Sawada, T., H. Tarui, N. Terada, M. Tanaka, T. Takahama, S. Tsuda and S. Nakano, *Theoretical analysis of textured thin-film solar cells and a guideline to achieving higher efficiency*, Proc. 23rd IEEE PV Specialists Conference, Louisville, KY, May (1993).

Schumm, G., *Chemical equilibrium description of stable and metastable defect structures in a-Si:H*, Phys. Rev. B **49** (1994) 2427-2442.

Selberherr, S., *Analysis and Simulation of Semiconductor Devices*, (Springer-Verlag, Wien, 1984).

Sichanugrist, P., M. Konagai, and K. Takahashi, *Theoretical analysis of amorphous silicon solar cells: Effects of interface recombination*, J. Appl. Phys. **55** (1984) 1155-1161.

Simmons J.G., and G.W. Taylor, *Nonequilibrium steady-state statistics and associated effects for insulators and semiconductors containing an arbitrary distribution of traps*, Phys. Rev. B **4** (1971) 502-511.

Shockley W., and W.T. Read, *Statistics of the recombinations of holes and electrons*, Phys. Rev. **87** (1952) 835-842.

Smole, F., and J. Furlan, *Effects of abrupt and graded a-Si:C:H/a-Si:H interface on internal properties and external characteristics of p-i-n solar cells*, J. Appl. Phys. **72** (1992) 5964-5969.

Smole, F., M. Topič, J. Furlan, *Amorphous silicon solar cell computer model incorporating the effects of TCO/a-Si:C:H junction*, Solar Energy Materials and Solar Cells **34** (1994) 385-392.

Stiebig, H., A. Kreisel, K. Winz, N. Schultz, C. Beneking, Th. Eickhoff and H. Wagner, *Spectral response modelling of a-Si:H solar cells using accurate light absorption profiles*, 1st World Conference on Photovoltaic Energy Conversion, (Proc. 24th IEEE PV Specialists Conference, Waikoloa, HI, USA, December 1994) 603-606.

Street R.A., *Hydrogenated amorphous silicon*, Cambridge Solid State Science Series, Eds. R.W. Cahn, E.A. Davis and I.M. Ward (Cambridge University Press, 1991).

Stutzmann, M., Phil. Mag. B **56** (1987) 63.

Stutzmann, M., *A comment on thermal defect creation in hydrogenated amorphous silicon*, Phil. Mag. Lett. **66** (1992) 147-150.

Swartz, G.A., *Computer model of amorphous silicon solar cell*, J. Appl. Phys. **53** (1982) 712-719.

Swartz, R.J., J.L. Gray, and G.B. Turner, *P-i-n thin film silicon hydrogenated alloy solar cells: numerical model predictions*, Technical Digest of the International PVSEC-1, Kobe, Japan (1984) 123-126.

Tasaki, H., W.Y. Kim, M. Hallerdt, M. Konagai, and K. Takahashi, *Computer simulation model of the effects of interface states on high performance amorphous silicon solar cells*, J. Appl. Phys. **63** (1988) 550-560.

Taylor, G.W., and J.G. Simmons, *Basic equations for statistics, recombination processes, and photoconductivity in amorphous insulators and semiconductors*, J. Non-Cryst. Solids **8-10** (1972) 940-946.

Tiedje, T., *Information about band-tail states from time-of-flight experiments*, in: Hydrogenated Amorphous Silicon, Part C (Semiconductors and Semimetals Vol. 21), Ed. Jacques I. Pankove (Academic Press, 1984) 207.

Wentinck, H.M., *Carrier injection in amorphous silicon devices*, Ph.D. thesis, Delft University of Technology, 1988.

Willemen, J.A., M. Zeman, and J.W. Metselaar, *Computer modeling of amorphous silicon tandem cells*, 1st World Conference on Photovoltaic Energy Conversion, (Proc. 24th IEEE PV Specialists Conference, Waikoloa, HI, USA, December 1994) 599-602.

Willemen, J.A., *Modeling of amorphous silicon single and multi-junction solar cells*, Ph.D. thesis, Delft University of technology, 1998.

Winer, K., *Defect formation in a-Si:H*, Phys. Rev. B **41** , (1990) 12150.

Zeman, M, G. Tao, M. Trijssenaar, J.A. Willemen, J.W. Metselaar, and R. Schropp, *Application of the defect pool model in modelling of a-Si:H solar cells*, in: Amorphous Silicon Technology - 1995, edited by M. Hack, E.A. Schiff, A. Madan, M. Powell, and A. Matsuda, Materials Research Society Symp. Proc. **377** (1995) 639-644.

Zeman, M., J.A. Willemen, L.L.A. Vosteen, G. Tao and J.W. Metselaar, *Computer modeling of current matching in a-Si:H/a-Si:H tandem solar cells on textured substrates*, Solar Energy Materials and Solar Cells **46** (1997) 81-99.

Zimmer, J., H. Stiebig, and H. Wagner, *Investigation of the electronic transport in PIN solar cells based on microcrystalline silicon by 2D numerical modeling*, in: Amorphous and Microcrystalline Silicon Technology - 1998, edited by R. Schropp, H. Branz, S. Wagner, M. Hack, and I. Shimizu, Materials Research Society Symp. Proc. **507** (1998) in print.

7 OPTICAL DEVICE MODELING

With such important advantages, interest in this approach (light trapping) has continued, but progress in this field has been hindered because there was no method available to calculate the degree of intensity enhancement to be expected.
—E. Yablonovitch and G.D. Cody, 1982

7.1 INTRODUCTION

A solar cell is a semiconductor optoelectronic device that is designed to work under illumination. According to Wenham and Green, 1996 one of the key attributes for achieving high-efficiency silicon solar cells is a good optical design, including means for achieving low surface reflection and high internal light trapping. Thus it is important to design a solar cell structure in which the absorption of incident light in the active parts of solar cell is maximized. In case of a-Si:H solar cells several light trapping techniques have been implemented to

The authors wish to thank Jože Furlan from the University of Ljubljana for his useful comments on this chapter.

achieve this aim. These include the introduction of textured (rough) surfaces and the use of special layers called reflectors in order to keep the light inside the active part of solar cell.

In order to achieve a good optical design accurate knowledge is required of the optical generation rate profile in the solar cell. The optical generation rate, G_{opt}, is also an important input parameter for electrical modeling. This rate is determined from the absorption profile of the photons in the cell. If it is assumed that every photon generates exactly one electron-hole pair the generation rate profile is equal to the absorption profile. The calculation of the absorption profile in a-Si:H based solar cells is not an easy task. Due to the implementation of various light trapping techniques these cells have become complex optical systems. Simple and straightforward analytical formulas for calculating the absorption that are based on the Lambert-Beer's absorption formula have been replaced by sophisticated optical models. In this approach the a-Si:H solar cell is regarded as a multilayer thin-film optical system and the optical behavior of this system, which has to take into account reflection and transmission at all interfaces and absorption in all layers of the system, is solved using numerical techniques.

In terms of optical systems, a-Si:H solar cells can be divided in two groups. The first group represents structures with smooth (flat) interfaces and the second group includes structures that contain one or more rough interfaces. The optical properties of a thin film multilayer optical system with flat interfaces can be calculated by using thin-film optics. The general treatment of optical properties of thin films can be found in several references, e.g., Heavens, 1955. This treatment uses the complex refractive indices of the media and the effective Fresnel's coefficients. Though all practical solar cells use textured substrates or superstrates and therefore their interfaces are rough, solar cell structures with flat interfaces are a very useful experimental tool for examining the optical models, for extracting unknown optical parameters, and for demonstrating the trends in optical behavior.

There is general consensus that the optical design of a-Si:H solar cells including light trapping techniques will play a major role in achieving higher efficiency. There is however still only limited understanding about the precise influence of interface texture on the optical absorption in the active layer of solar cell. Optical modeling, which comprises the development of new optical models that include scattering, is gaining more interest, because the results lead to better insight in the optical behavior of solar cell structures with rough interfaces.

The texture causes scattering of light incident at the interface and in general, the amount and the angular distribution of scattered light depend on the refractive indices of the media, the texture of the interface, and the incident angle. If the exact morphology of the rough interfaces is known, one can apply several approaches such as geometrical optics, physical optics or electromagnetic theory to study the scattering of light at these interfaces. As the texture introduces spatial variations in all three dimensions of device structure, the precise optical modeling of the solar cells on textured substrates should be carried out using at least two-dimensional (2-D) modeling. The 2-D modeling was applied to study the effects of scattering in a-Si:H solar cells only for simple, regular texture patterns (Furlan et al., 1994, Sawada et al., 1994). However, since the morphology of the textured interfaces in a-Si:H solar cells is usually random and the typical dimensions of interface features are of the same order of magnitude as the wavelength of light, the exact calculation of the scattering effects in solar cells, whether 1-D or 2-D, is still complicated and requires long computational time. Instead of exact calculations, semi-empirical 1-D optical models based on thin-film optics have been developed (Tao et al., 1994, Leblanc et al., 1994, Stiebig et al., 1994). These models use the average scattering data of the rough interfaces in a-Si:H solar cells that can be determined experimentally.

Scattering data of textured superstrates such as the textured TCO layer on glass is available mainly for the TCO/air interface. One can expect that these data do not accurately describe the actual scattering at the TCO/a-Si:H interface. There is almost no scattering data available for textured interfaces within the a-Si:H based solar cell. Nevertheless, numerical simulations using advanced optical models including scattering have already contributed to the understanding of the optical behavior of a-Si:H solar cells. Numerical simulations, combined with measured data such as the total reflectance and spectral response of the solar cell, can be used to extract the scattering data (Van den Berg et al., 1998). Still a lot of theoretical, experimental and modeling work has to be conducted in order to fully understand these matters so that eventually efficient optical optimization of a-Si:H solar cell structure is possible.

7.2 OPTICAL PROPERTIES OF THE LAYERS AND INTERFACES IN a-Si:H SOLAR CELL

The optical properties of all individual layers are described by the complex refractive index $\tilde{n} = n - ik$. The real part n is called the refractive index, and the imaginary part k is called the extinction coefficient. The extinction coefficient is related to the absorption coefficient α by:

$$k = \frac{\alpha\lambda}{4\pi} \tag{7.1}$$

The quantities n and k are often called optical constants, although they are functions of the wavelength of the light or the photon energy. The wavelength λ is related to the photon energy $h\nu$ as

$$h\nu = \frac{h}{q}\frac{c}{\lambda} \tag{7.2}$$

where the photon energy is expressed in eV, h is the Planck's constant, q is the charge of an electron, and c is the velocity of light in vacuum. In case of a-Si:H solar cells it is sufficient to know the optical constants of the layers in the wavelength range from 300 nm to 1200 nm (or expressed as photon energies from 4.1 eV to 1.0 eV).

The optical data of a-Si:H alloy layers above their optical gap are usually obtained from reflection and transmission (R-T) measurements. In this measurement the reflectance R and the transmittance T of a thin film deposited on transparent flat substrates such as glass or quartz is measured. The refractive index, the extinction coefficient and the thickness of the film are obtained from the measured R and T by fitting the calculated R and T to the measured data. A good overview of this procedure is given in (Ley, 1984, Van den Heuvel, 1989). In order to determine the optical constants in the low absorption region, i.e., below the optical band gap, where the R-T measurement is not accurate, several experimental techniques are used such as the Constant Photocurrent Method (CPM) (Moddel et al., 1980) and Photothermal Deflection Spectroscopy (PDS) (Curtins and Favre, 1988).

The optical properties of metals that are used as the electrodes such as aluminium or silver are readily found in the handbook literature. On the other hand, the optical properties of TCOs are not easily obtained. The most common TCO that is used in a-Si:H solar cells on glass substrate is SnO_2:F. As this coating is textured and usually comprises a compositional gradient, it is difficult to determine its optical properties in a straightforward way. The optical constants of the layers which comprise a typical superstrate single junction a-Si:H solar cell are shown in Fig. 7.1.

The texture of the superstrate is introduced to all subsequent interfaces in the solar cell. In order to describe the correlation between the rough interface and its scattering properties, the following parameters are important: the root mean square (*rms*) roughness σ_r, and the *rms* slope m of the textured surface, the scattering data of the rough interfaces, which include the haze parameter and the angular distribution of scattered light.

The *rms* roughness is derived from the "height distribution function" that describes the height of rough surface above and below the mean surface level. Experimentally, it is found that in nearly all cases the height distribution func-

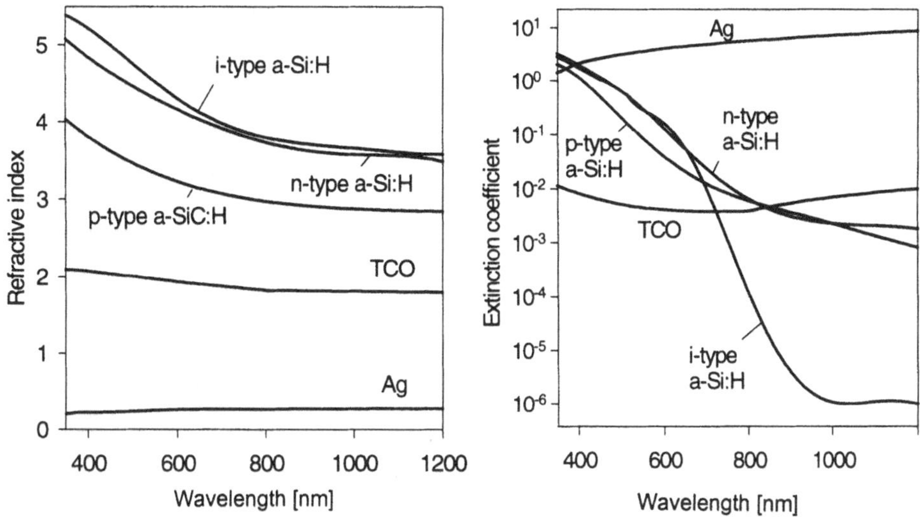

Figure 7.1. The refractive index and the extinction coefficient of individual layers that comprise a typical superstrate single junction a-Si:H solar cell.

tions are Gaussian to a very good approximation (Wallinga, 1998). The *rms* roughness is the variance of the Gaussian height distribution function. In a similar way the *rms* slope is defined from the Gaussian "slope distribution function" of the rough surface. Scanning electron microscopy (SEM), transmission electron microscopy (TEM), or atomic force microscopy (AFM) are widely used to estimate the roughness of the surfaces. Fig. 4.3 shows a SEM photograph of an Asahi U-type textured tin oxide substrate and Fig. 4.4 shows the surface of a typical 500 nm thick complete a-Si:H solar cell deposited on this substrate. It is evident that the texture of the tin oxide substrate is smoothened by the deposition of amorphous silicon.

When light is incident on a rough interface, a part of the light is reflected and transmitted in the specular direction, and the other part is diffusely reflected and transmitted. The ratio of the diffused reflectance to the total reflectance and the diffused transmitance to the total transmitance is called the haze value of the reflectance and transmitance, respectively. The haze value can be defined for a monochromatic light, usually at 700 nm, or over a wavelength interval. Asahi Glass Company reports the haze value of their textured substrate over the interval between 400 nm to 800 nm (Mizuhashi et al., 1988). Angle-resolved spectroscopy (ARS) both for the reflectance and transmittance is used for obtaining information about the haze and angular distribution of the scattered light (Leblanc et al., 1994, Tao et al., 1994, Wallinga, 1998).

The measurement of the external parameters of a solar cell such as the open circuit voltage, the short circuit current, the fill factor, and the efficiency is carried out under standard conditions. The illumination spectrum used in this measurement is the AM1.5 solar spectrum normalized to a power intensity of 100 mW/cm^2. Therefore, the absorption profiles in the solar cells are calculated using the same spectrum. The spectral irradiance data for this spectrum are published by Hulstrom et al., 1985.

7.3 LAMBERT-BEER'S FORMULA

The most frequently used approach for the calculation of the absorption profile of photons in semiconductor devices makes use of Lambert-Beer's absorption formula. This formula states that a photon flux density after passing a distance x in a film with an absorption coefficient $\alpha(\lambda)$ is reduced with a factor $e^{-\alpha(\lambda)x}$

$$\Phi(x,\lambda) = \Phi^0(0,\lambda)\exp(-\alpha(\lambda)x) \tag{7.3}$$

where $\Phi^0(0,\lambda)$ is the incident photon flux density. The photon flux density is the number of photons per unit area per unit time and per unit wavelength. It is related to the power density $P(\lambda)$ associated with the solar radiation (also called irradiance) by

$$\Phi^0(\lambda) = P(\lambda)\frac{\lambda}{hc} \tag{7.4}$$

The spectral generation rate g_{sp}, which is the number of electron-hole pairs generated per second per unit volume per unit wavelength, at a depth x in the film by photons with a wavelength λ is calculated using the following formula, on the assumption of zero reflection,

$$g_{sp}(x,\lambda) = \eta_g\,\Phi^0(\lambda)\,\alpha(\lambda)\,e^{-\alpha(\lambda)x} \tag{7.5}$$

where η_g is the generation quantum efficiency, usually assumed unity. This assumption implies that every photon generates exactly one electron-hole pair. The optical generation rate G_{opt} is calculated from the spectral generation rate by integrating over the appropriate wavelength range,

$$G_{opt}(x) = \int_{\lambda_1}^{\lambda_2} g_{sp}(x,\lambda)\,d\lambda \tag{7.6}$$

The optical generation rate is related to the absorption profile $A(x)$ in the film:

$$G_{opt}(x) = \eta_g A(x) \tag{7.7}$$

In order to take into account the reflections from the front surface and the back contact, and the absorption in the non-active parts of the solar cell, such as the glass substrate and the TCO, the simple expression for calculating the spectral generation rate (Eqn. 7.5) in the active layer of a-Si:H solar cell has been extended (Schade and Smith, 1985b, Block, 1993, Bruns, 1993). The resulting formula for the spectral generation rate for photons of wavelength λ is

$$g_{sp}(x) = \eta_g \, \Phi^0 \, (1 - R_f) \, e^{-\alpha_{gl} d_{gl}} \, e^{-\alpha_{\mathrm{TCO}} d_{\mathrm{TCO}}} \alpha \, \left(e^{-\alpha x} + R_b \, e^{-2\alpha d_i} \, e^{\alpha x}\right) \tag{7.8}$$

where R_f is the reflectance from the front side, R_b is the reflectance from the back contact, d_i is the thickness of the active part of the solar cell, α_{gl} (α_{TCO}) is the absorption coefficient of the glass substrate (of the TCO layer), and d_{gl} (d_{TCO}) is the thickness of the glass substrate (of the TCO layer). With the data for the above mentioned quantities as a function of wavelength it is easy to calculate the optical generation rate profile in a solar cell using Eqn. 7.6. The formula can be further extended in order to take account the absorption in the p^+- and n^+-type layer.

In a device with flat interfaces only, interference occurs between the reflected and incident light when the wavelength of the light is comparable to the thickness of the cell. In the approach using Lambert-Beer's formula, this interference is neglected. Instead, in the case of rough interfaces the effects of scattering are taken into account. Schade and Smith, 1985b derived an analytical expression for the total absorptance of photons in the active layer of a-Si:H solar cell including the effects of scattering. They assumed that the scattering interfaces had random roughness and that the angular distribution of the scattered light was Lambertian ($\Phi(\theta) = \Phi^0 \cos\theta$). The first effect of light scattering on the film absorptance originates from the fact that light traveling through a film of thickness w under an angle θ with respect to the normal has an effective path length of $w/\cos\theta$. The effective absorptance in the film is given by

$$A_{\mathit{eff}} = \int_{\theta=0}^{\pi/2} A(\theta) \frac{dP(\theta)}{P_i} \tag{7.9}$$

where

$$A(\theta) = (1 - \exp(-\alpha w / \cos\theta)) \tag{7.10}$$

and $dP(\theta)/P_i$ is the fraction of the incident power P_i that is scattered within the angle θ, i.e.,

$$\frac{dP(\theta)}{P_i} = 2\sin(\theta)\cos(\theta)\, d\theta \tag{7.11}$$

For weakly absorbed light, these expressions yield $A_{eff} \approx 2\alpha w$, about twice the absorptance of specularly transmitted light. If the interfaces are not completely scattering, the absorptance of the film is weighted according to the fraction of scattered light S, so that

$$A_{eff} \approx S\,(2\alpha w) + (1 - S)\,(\alpha w) \tag{7.12}$$

The fractions of incident (specular) light that are scattered at the front and back interfaces are S_f and S_b, respectively. For the first pass of light from the front to the back contact $S = S_1 = S_f$. For the light reflected from the back contact and fractionally scattered $S = S_2 = S_f + (1 - S_f)\,S_b$.

Light reflected from the rear contact can either escape through the front of the cell or be trapped by internal reflection if $\theta > \theta_c$, where θ_c is the critical angle. For a multilayer system, such as a-Si:H/TCO/glass/air, with indices of refraction $n_{\text{a-Si:H}} > n_{\text{TCO}} > n_{gl} > n_{air}$, the critical angle is given by:

$$\sin\theta_c = n_{air}/n_{\text{a-Si:H}} \tag{7.13}$$

independent of n_{TCO} and n_{gl}. If the light has a Lambertian distribution the fraction of energy that is trapped by total internal reflection R_t is

$$R_t = 1 - \sin^2\theta_c \tag{7.14}$$

In this way light can undergo several passes through the cell. In the limiting case when $S_f = S_b = 1$ and $\alpha d < 1$, the total absorptance after multiple passes of light through the cell is (Schade and Smith, 1985b)

$$A_{eff} = 2\alpha d \frac{1 + (1 - 2\alpha d)\,T_n^2 R_b}{1 - (1 - 2\alpha d)^2\,T_n^2 R_b T_p^2 R_t} \tag{7.15}$$

where T_n and T_p are the transmittances of the n and p layer, respectively.

Schade and Smith, 1985a used a simplified scattering mechanism by assuming that the scattering centers act independently so that Mie scattering theory could be applied to evaluate the fractions of ligth scattered at the rough interfaces. They measured the forward and backscattered fractions of incident light by rough SnO_2:F films on glass substrates and the backscattered fractions of light by such films overcoated with Ag and compared them to the model. These results were used in calculations of the quantum efficiency of a-Si:H solar cells on rough SnO_2:F TCO substrates (Schade and Smith, 1985b). After calibrating the model parameters by comparing the measured quantum efficiency of a p-i-n a-Si:H solar cell with the calculated results, the effect of TCO roughness and thickness, optical gap and thickness of the p-type layer, the back contact reflectivity, and the thickness of the i-layer on the quantum efficiency was simulated. The authors determined the trends in the optical behavior of the solar cell as a function of these parameters and found good agreement between the observed and calculated properties.

7.4 OPTICAL SYSTEM WITH SMOOTH INTERFACES

7.4.1 Multilayer optical system

A more precise approach for determination of the absorption profile in a solar cell is to treat the solar cell as a multilayer optical system and to calculate the optical behavior of this system (reflectance, transmittance and absorptance) using the theory of thin film optics. The reflection and transmission at all interfaces and the absorption in all layers of the system are then taken into account. Here, we discuss the optical properties of a multilayer system with flat interfaces.

Reflection and transmission at a flat interface. The simplest optical system is a flat optical interface, which is the boundary between two optical media, each described by its complex refractive index $\tilde{n} = n - ik$. When light impinges on the optical interface, a part of the light is reflected from the interface and the other part is transmitted through it. The reflectance is the ratio of the energy reflected at the surface or an interface to the incident energy and the transmittance is the ratio of the transmitted energy to the incident energy. The flux in the direction of incidence is referred to as the positive-going and the flux in the opposite direction as the negative-going flux.

For light arriving at the optical interface at the normal incidence, the reflectance and transmittance can be calculated from the reflection and transmission coefficients known as the Fresnel amplitude coefficients ($\tilde{r}$, $\tilde{t}$) (Bennett and Bennett, 1978)

$$\tilde{r} = \frac{\tilde{n}_0 - \tilde{n}_1}{\tilde{n}_0 + \tilde{n}_1} \tag{7.16}$$

$$\tilde{t} = \frac{2\tilde{n}_0}{\tilde{n}_0 + \tilde{n}_1} \tag{7.17}$$

The $\tilde{n}_0$ and $\tilde{n}_1$ are the complex refractive indices of the first and second medium, respectively. The Fresnel amplitude coefficients ($\tilde{r}$, $\tilde{t}$) are also complex numbers. The reflectance and transmittance of the interface are

$$R = |\tilde{r}|^2 = \left| \frac{\tilde{n}_0 - \tilde{n}_1}{\tilde{n}_0 + \tilde{n}_1} \right|^2 \tag{7.18}$$

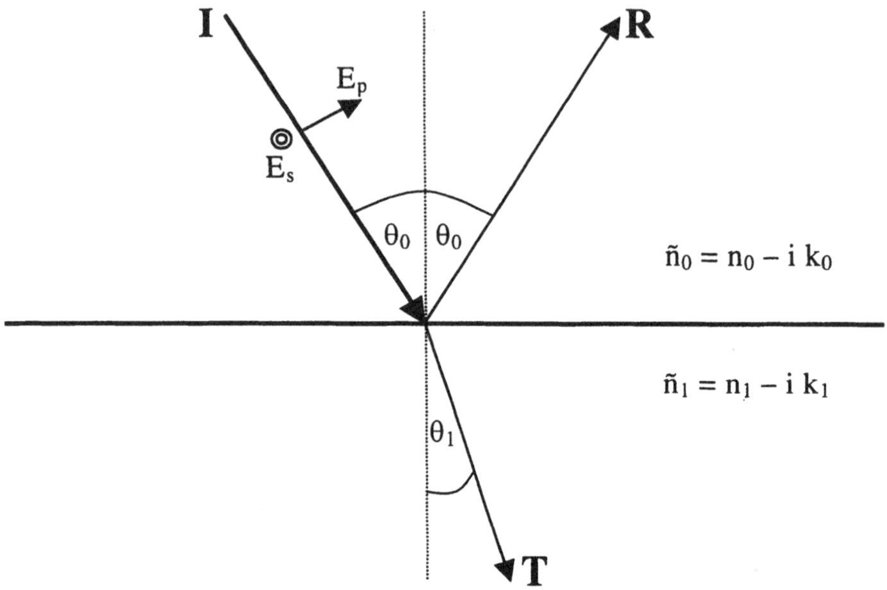

Figure 7.2. The reflection and transmission of light at a flat interface.

$$T = \left|\frac{\tilde{n}_1}{\tilde{n}_0}\right| \; |\tilde{t}|^2 = \frac{4\;|\tilde{n}_0\,\tilde{n}_1|}{|\tilde{n}_0 + \tilde{n}_1|^2} \tag{7.19}$$

When the incident angle of light is unequal to zero (i.e., not the case of normal incidence), see Fig. 7.2, the reflectance and transmittance depend on the angle of incidence and the polarization of the light. Let us denote θ_0 as the angle of incidence and θ_1 as the angle of refraction. In non-absorbing media the angles are real numbers and satisfy Snell's law

$$\frac{n_0}{n_1} = \frac{\sin(\theta_1)}{\sin(\theta_0)} \tag{7.20}$$

When the media are absorbing, complex values of refractive indices ($\tilde{n}_0$, $\tilde{n}_1$) and angles (θ_0, θ_1) should be used, which then satisfy Snell's law with complex values.

The polarization of light is defined as follows: the electric field component that oscillates perpendicular to the plane of incidence is called the s-component E_s. The component in the plane of incidence is called the p-component E_p. The plane of incidence is the plane which is defined by the incident ray and the normal to the interface. The reflectance and transmittance can still be calculated by the Fresnel amplitude coefficients using effective refractive indices for s-polarized and p-polarized light (Bennett and Bennett, 1978). For the s-polarized light the effective refractive index is calculated as

$$\tilde{n}_{is,eff} = \tilde{n}_i \cos\Theta_i \tag{7.21}$$

for i = 0, 1. The Fresnel amplitude coefficients are

$$\tilde{r}_s = \frac{\tilde{n}_{0s,eff} - \tilde{n}_{1s,eff}}{\tilde{n}_{0s,eff} + \tilde{n}_{1s,eff}} \tag{7.22}$$

$$\tilde{t}_s = \frac{2\tilde{n}_{0s,eff}}{\tilde{n}_{0s,eff} + \tilde{n}_{1s,eff}} \tag{7.23}$$

The reflectance and transmittance for s-polarized light are:

$$R_s = |\tilde{r}_s|^2 \tag{7.24}$$

$$T_s = \left|\frac{\tilde{n}_{1s,eff}}{\tilde{n}_{0s,eff}}\right| \; |\tilde{t}_s|^2 \tag{7.25}$$

For p-polarized light the effective refractive index is calculated as

$$\tilde{n}_{ip,eff} = \frac{\tilde{n}_i}{\cos\Theta_i} \tag{7.26}$$

for i = 0, 1. The Fresnel amplitude coefficients are

$$\tilde{r}_p = \frac{\tilde{n}_{0p,eff} - \tilde{n}_{1p,eff}}{\tilde{n}_{0p,eff} + \tilde{n}_{1p,eff}} \tag{7.27}$$

$$\tilde{t}_p = \frac{2\tilde{n}_{0p,eff}}{\tilde{n}_{0p,eff} + \tilde{n}_{1p,eff}} \tag{7.28}$$

The reflectance and transmittance for p-polarized light are

$$R_p = |\tilde{r}_p|^2 \tag{7.29}$$

$$T_p = \left|\frac{\tilde{n}_{1p,eff}}{\tilde{n}_{0p,eff}}\right| \; |\tilde{t}_p|^2 \tag{7.30}$$

The resulting reflectance and transmittance of the interface is usually calculated as the mean value of the two polarized components

$$R = \frac{R_s + R_p}{2} \tag{7.31}$$

$$T = \frac{T_s + T_p}{2} \tag{7.32}$$

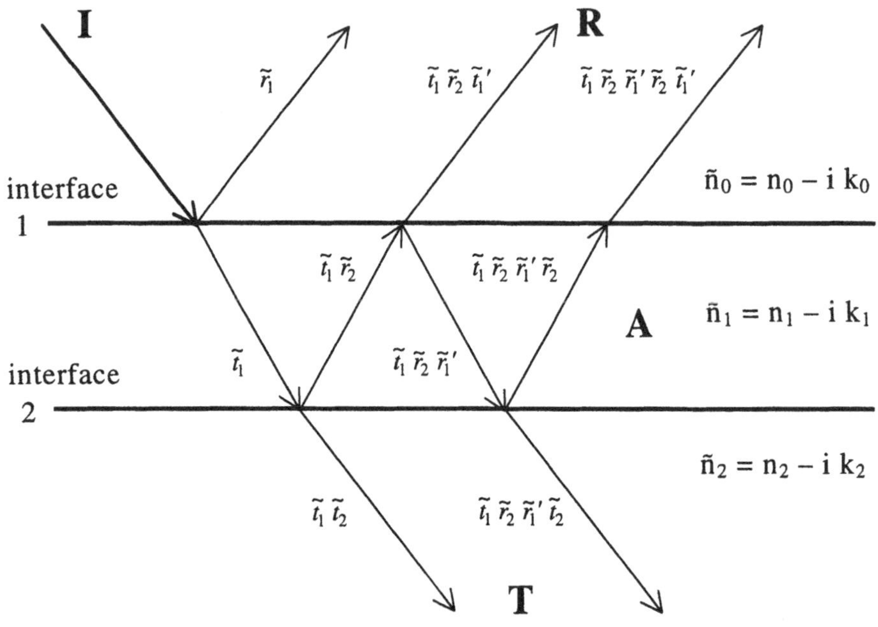

Figure 7.3. The reflection and transmission of light in a single layer.

Single layer with two flat interfaces. The optical system of a single layer consists of two interfaces and the thickness of the layer. At both interfaces reflection and transmission takes place and the slab causes a phase shift and light absorption. Multiple reflections of the light can occur between the two interfaces which complicate the calculation of the reflectance and transmittance as illustrated in Fig. 7.3. For this system we can distinguish the two cases: the layer can be considered optically coherent for a particular light beam or optically incoherent. A layer is considered to be coherent if its thickness is much smaller than the coherence length of the light beam that is incident on the system. The coherence length is determined by the type of light source and the bandwidth of the light, as produced, eg., by a monochromator. The coherence length of the light emitted by a thermal light source, such as tungsten filament lamp, is in the order of a millimeter (Milonni and Eberly, 1988). The coherence length corresponding to the bandwidth $\Delta\lambda$ of a monochromator is determined by (Ley, 1984)

$$l_c = \frac{\lambda^2}{\Delta\lambda}\frac{1}{2\pi} \tag{7.33}$$

A typical value for $\Delta\lambda$ is 4 nm which gives a coherence length of 40 μm at λ= 1 μm. The coherence length is determined by the smaller value of the coherence lengths obtained for the light source and the monochromator, i.e. 40 μm.

In the following we consider the case of normal incidence and a coherent layer. The reflectance and transmittance of a single layer can be calculated by using the effective Fresnel amplitude coefficients $\tilde{r}$ and $\tilde{t}$

$$\tilde{r} = \tilde{r}_1 + \frac{\tilde{t}_1\tilde{t}_1'\,\tilde{r}_2\,e^{-2i\tilde{\delta}_1}}{1-\tilde{r}_2\,\tilde{r}_1'\,e^{-2i\tilde{\delta}_1}} \tag{7.34}$$

$$\tilde{t} = \frac{\tilde{t}_1\tilde{t}_2\,e^{-i\tilde{\delta}_1}}{1-\tilde{r}_2\,\tilde{r}_1'\,e^{-2i\tilde{\delta}_1}} \tag{7.35}$$

$$\tilde{\delta}_1 = \frac{2\pi}{\lambda}d_1\tilde{n}_1 = \frac{2\pi}{\lambda}d_1\,(n_1 - ik_1) \tag{7.36}$$

where $\tilde{\delta}_1$ is the complex phase shift of light and d_1 the thickness of the layer. The real part of $\tilde{\delta}_1$ represents the shift in the phase of the light wave and the imaginary part represents the absorption due to one pass of the light passes through the thickness of the layer. The $\tilde{r}_i$, $\tilde{t}_i$, $\tilde{r}_i'$ and $\tilde{t}_i'$ are the Fresnel coefficients of the individual interface i (i = 1, 2). The coefficients $\tilde{r}_i$ and $\tilde{t}_i$ represent the Fresnel coefficients for positive-going flux and $\tilde{r}_i'$ and $\tilde{t}_i'$ represent Fresnel coefficients for the negative-going flux. Using *effective* Fresnel coefficients $\tilde{r}$ and $\tilde{t}$, the single layer can be regarded as one effective interface, which is described by the reflectance R, transmittance T, and the total absorptance A.

$$R = |\tilde{r}|^2 \tag{7.37}$$

$$T = \left|\frac{\tilde{n}_2}{\tilde{n}_0}\right|\;|\tilde{t}|^2 \tag{7.38}$$

$$A = 1 - R - T \tag{7.39}$$

The normalized absorption profile in the layer is given by:

$$A'\,(x) = \alpha_1\left|\frac{\tilde{n}_1}{\tilde{n}_0}\right|\;|\tilde{t}_1|^2\left|\frac{e^{-i\tilde{\delta}_1 x/d_1} + \tilde{r}_2\,e^{-i\tilde{\delta}_1}\;e^{-i\tilde{\delta}_1(d_1-x)/d_1}}{1-\tilde{r}_2\,\tilde{r}_1'\;e^{-2i\tilde{\delta}_1}}\right|^2 \tag{7.40}$$

In the case of an incoherent layer, there is no interference effect and the reflectance and transmittance are calculated by adding up the energies (i.e. the square of the amplitudes) instead of the amplitudes. The reflectance, transmittance, absorptance, and the normalized absorption profile are as follows

$$R = R_1 + \frac{T_1 R_2 T_1' \, e^{-2\alpha_1 d_1}}{1 - R_2 \, R_1' \, e^{-2\alpha_1 d_1}} = \left|\tilde{r}_1\right|^2 + \frac{\left|\tilde{t}_1\right|^2 \, \left|\tilde{r}_2\right|^2 \, \left|\tilde{t}_1'\right|^2 e^{-2\alpha_1 d_1}}{1 - \left|\tilde{r}_2\right|^2 \, \left|\tilde{r}_1'\right|^2 \, e^{-2\alpha_1 d_1}} \tag{7.41}$$

$$T = \left|\frac{\tilde{n}_2}{\tilde{n}_0}\right| \frac{T_1 T_2 \, e^{-\alpha_1 d_1}}{1 - R_2 \, R_1' \, e^{-2\alpha_1 d_1}} = \left|\frac{\tilde{n}_2}{\tilde{n}_0}\right| \frac{\left|\tilde{t}_1\right|^2 \, \left|\tilde{t}_2\right|^2 e^{-\alpha_1 d_1}}{1 - \left|\tilde{r}_2\right|^2 \, \left|\tilde{r}_1'\right|^2 \, e^{-2\alpha_1 d_1}} \tag{7.42}$$

$$A = 1 - R - T \tag{7.43}$$

$$A'(x) = \alpha_1 \left|\frac{\tilde{n}_1}{\tilde{n}_0}\right| \left|\tilde{t}_1\right|^2 \frac{e^{-\alpha_1 x} + \left|\tilde{r}_2\right|^2 \, e^{-\alpha_1 (2d_1 - x)}}{1 - \left|\tilde{r}_2\right|^2 \, \left|\tilde{r}_1'\right|^2 \, e^{-2\alpha_1 d_1}} \tag{7.44}$$

Multilayer system with flat interfaces. A coherent single layer can be described by the effective Fresnel coefficients $\tilde{r}$ and $\tilde{t}$ (Eqns. 7.34 and 7.35). In this way the coherent single layer can be considered as one effective interface. In the case of a multilayer structure we iteratively apply the same procedure of describing a layer as an effective interface to all subsequent layers of the system. The whole multilayer structure is thus described as one effective interface and the reflectance, transmittance and absorptance can be calculated from Eqns. 7.37 to 7.39.

For a multilayer structure that consists of incoherent layers, the reflectance and transmittance can be calculated iteratively using Eqns. 7.41 and 7.42. When the multilayer structure contains both coherent and incoherent layers, the coherent layers are treated first and then the incoherent layers.

7.5 OPTICAL SYSTEM WITH ROUGH INTERFACES

7.5.1 *Light trapping*

In order to increase the absorption of the incident light in a:Si:H solar cells light trapping techniques are implemented. The term light trapping is used to describe methods to capture the light in a desired part of the solar cell and preventing it from escaping. The techniques used to keep the light inside the solar cell include the introduction of textured substrates and special back reflectors. The utilization of textured substrates leads to a large suppression of the reflection optical loss at the front of the cell and the implementation of

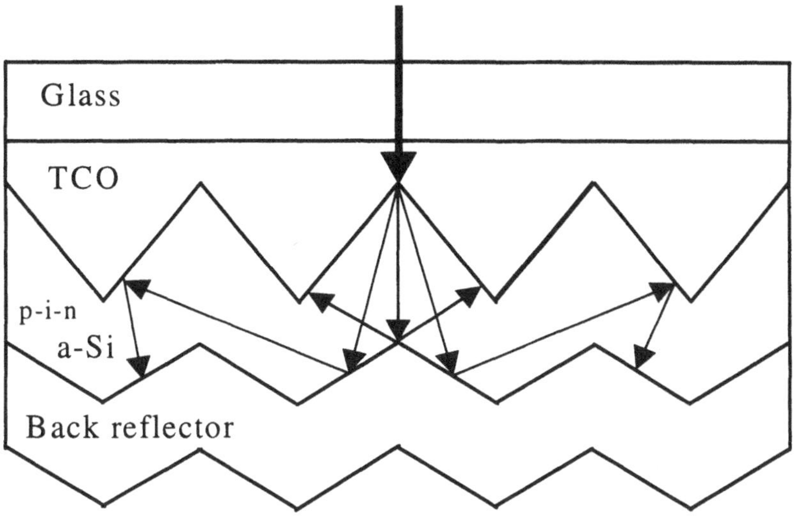

Figure 7.4. Schematic representation of light trapping in an a-Si:H solar cell.

back reflectors minimizes the transmission loss into the back contact of the cell. A schematic representation of light trapping in an a-Si:H solar cell is shown in Fig. 7.4. The incident light passes through the glass and the TCO coating and arrives at the textured interface with the silicon. Due to the texture a fraction of light is scattered into the silicon layer under various angles. The light that is not absorbed in the silicon layer arrives at the back contact. The light that is transmitted into the back metal contact is an optical loss and therefore the reflectance of the silicon/back contact interface has to be maximized. This is achieved by adding a thin TCO layer between the n-type a-Si:H and the Al or Ag layer. The TCO and metal layer then form the back contact. The interface between the silicon and the back contact is also rough and causes scattering of a part of the light that is reflected at this interface. The reflected light passes through the silicon back to the TCO/a-Si:H interface. Since the refractive index of a-Si:H is higher than the refractive index of TCO, the fraction of the light arriving at this interface from the a-Si:H side under an angle greater than the critical angle is reflected back into the silicon layer. The light can thus pass several times through the a-Si:H layer. The combination of the internal reflection at the TCO/a-Si:H interface with the highly reflective back contact leads to light trapping in the a-Si:H layer.

Light trapping in highly efficient a-Si:H solar cells has become a standard technique. This development has urged the need for sophisticated optical mod-

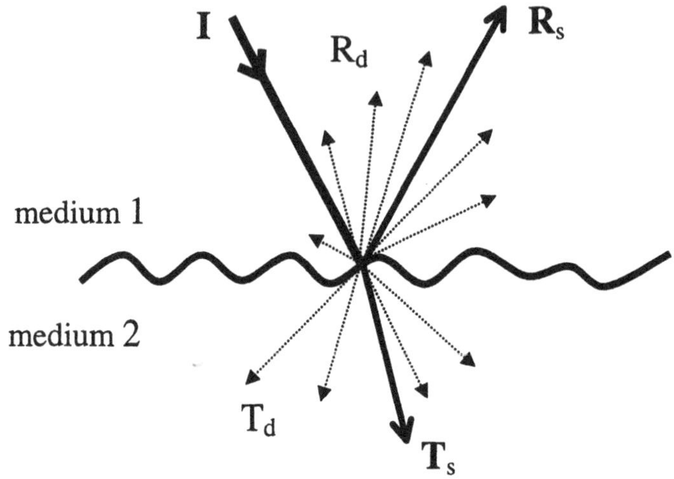

Figure 7.5. The reflection and transmission processes at a rough interface.

els which are able to accurately calculate the optical behavior of a-Si:H solar cells on textured substrates.

7.5.2 Scattering at a rough interface

The principle of light absorption enhancement by employing a textured substrate is to take advantage of light scattering at a rough interface. When light reaches a rough interface, a part of it will be scattered in various directions instead of propagating in the specular direction. In this way the average light path in a layer is increased and the light absorption enhanced. The scattered part of light is also called diffuse light. Fig. 7.5 illustrates the processes of scattering when light is incident on a rough interface. A part of the incident light will be reflected and transmitted in the specular direction, the remainder will be diffusely reflected and transmitted. We denote the specular reflectance as R_s, the specular transmittance in the refracted direction as T_s, the diffuse reflectance and transmittance as R_d and T_d, respectively. Energy conservation requires that

$$R_s + T_s + \sum R_d + \sum T_d = 1 \tag{7.45}$$

Yablonovitch and Cody, 1982 have shown that the effective absorption of a textured semiconductor sheet can be enhanced as much as a factor $4n^2$, where n is the refractive index of the semiconductor. This theoretical maximum is reached when the texture of the interface where the light enters the sheet scatters the photons with a fully randomized angular distribution. A fraction of these photons is trapped by total internal reflection. This result is contained in Eqn. 7.15 as a special case for $R_b = T_n = T_p = 1$, which yields (Schade and Smith, 1985b)

$$A_{eff} \cong 4\alpha d \left(n_{\text{a-Si:H}}/n_{air}\right)^2 \tag{7.46}$$

The first application of a textured interface in a-Si:H solar cell for optical absorption enhancement showed an increase of AM1 short circuit current by 3.0 mA/cm^2 (Deckman et al., 1983).

Several groups have investigated the relation between the average grain size on the surface of SnO_2 and the external parameters of a-Si:H solar cells. Iida et al., 1983 observed that the specular reflectivity from the glass substrate was reduced to 7 % with an increase of the grain size from 0.1 to 0.8 μm, and the short circuit current was increased from 12 to 14 mA/cm^2. He also studied the effect of the texture on the collection efficiency of the solar cell and demonstrated that texture at the incident side improved both the short (300 to 600 nm) and long (600 to 800 nm) wavelength response, while texturing at the back side improved only the long wavelength response (Iida et al., 1987). He concluded that the SnO_2 film with an average grain size of around 0.2 μm and larger was effective for sufficient optical confinement of the incident light over a wide wavelength range. Gordon et al., 1989 systematically varied the amount of texturing of the tin oxide by increasing the film thickness and his investigation revealed that only 5 to 10 % diffuse scattering is necessary to obtain the entire beneficial effect of light trapping: higher values gave rise to poorer cell performance. Hegedus and Buchanan, 1996 investigated several commercially available TCO's (SnO_2 and ZnO) with different haze values and their influence on the current generation in a-Si:H solar cells. Their results show that haze values in excess of 5 % had only limited effect on current generation at long wavelengths. They observed that for a single junction solar cell, using recently available textured ZnO, the current generation is about 0.6 mA/cm^2 larger than using textured SnO_2.

However, the advantage of using ZnO as a front TCO to obtain higher J_{sc} must be balanced with the apparent difficulty in maintaining V_{oc} and fill factor.

When working with rough interfaces it is of major importance to know the ratio between the fractions of specular and diffused light and the angular distribution of the diffused light. The theoretical treatment of scattering depends on the morphology of the rough interface. In the case of solar cells the texture of the interfaces replicates the shape of the irregularities introduced by the substrate. The size of the irregularities is in order of a few tenths of a micrometer (see Fig. 4.3), comparable to the wavelength of incident light in the medium (a-Si:H). Two cases are distinguished: i) irregularities that can be considered as isolated particles and ii) irregularities that are so closely spaced that they cannot be treated as independent scattering centers. In the second case the effect of each center is correlated with that of its neighbors and scattering from such ensembles is often called "microirregularity scattering" (Elson et al., 1979). When the scattering centers can be considered to act independently, Mie scattering theory can be used to evaluate scattering from these centers. The scattering center is then described as a spheral particle, characterized by a radius, r. This approach was used by Schade and Smith, 1985a in order to determine the fractions of light that are forward scattered and backscattered by rough SnO_2:F films on glass substrates and backscattered by such films overcoated with Ag (see Section 7.3).

When a rough surface consists of correlated scattering centers it can be regarded as a flat interface with a small stochastic disturbance, which is characterized by the *rms* roughness σ_r. The scalar theory that is based on the solution of the Helmholtz-Kirchhof diffraction integral is applied to such surfaces. The scalar theory was developed for surfaces whose *rms* roughness is much less than the wavelength of the light (Davies, 1954). Later, the theory was extended for surfaces whose *rms* roughness is comparable with the wavelength of the light (Porteus, 1963). Bennett and Porteus, 1961 demonstrated that the specular fraction of the total reflectance is related to the *rms* roughness σ_r and the incident angle θ_0 by

$$\frac{R_s}{R_s + R_d} = \exp\left[-\left(\frac{4\pi\sigma_r n_0 \cos(\theta_0)}{\lambda}\right)^2\right] \tag{7.47}$$

where n_0 is the refractive index of the medium of incidence.

The size and shape of the irregularities of the textured interfaces in a-Si:H solar cells have a broad distribution and therefore the exact calculations, which should ideally be 3-D, are complicated and would require enormous computational facilities. Further, the size scale of the irregularities is in the same order as the wavelength of the incident light, which complicates the theoretical treatment, as neither the Mie theory nor the Rayleigh approach can be readily applied. Only for simple regular texture patterns 2-D modelling was applied to study the effects of texture. Furlan et al., 1994 used geometrical optics for V-grooved device structures and Sawada et al., 1994 developed a 2-D program implementing a model in which the Helmholtz equation is solved for rounded periodic gratings. The calculations of Sawada et al., 1994 show that in case of normal incidence the highest generation rate is at the bottom in the valleys of the periodic grating formed by the TCO. Similarly, Furlan et al., 1994 reports that the maximum generation rate is at the bottom of the grooves in the TCO.

Several semi-empirical, 1-D models exist in the literature, which take into account the effect of textured substrates and the back reflectors (Deckman et al.,

1983, Schade and Smith, 1985a, Morris et al., 1990, Leblanc et al., 1994, Tao et al., 1994, Stiebig et al., 1994). The models introduced by Deckman et al., 1983, Schade and Smith, 1985a, and Morris et al., 1990 are based on the following assumptions: (i) absorption is calculated using Lambert-Beer's formula, by which light coherence is not taken into account; (ii) the diffuse reflectance and transmittance of a rough interface are either experimentally determined or derived from the Mie theory and the angular distribution of scattered light is assumed to be Lambertian, (iii) only two interfaces are considered to be rough, the TCO/p^+interface and the n^+/back reflector interface. This assumption is justified if there is only a small difference in refractive index between the various layers within the p-i-n structure. Using these assumptions the effect of rough interfaces is taken into account by introducing the increased effective path length of light through the intrinsic layer and the reduced front reflection. The general features of these models are presented in Section 7.3.

The models reported by Leblanc et al., 1994, Stiebig et al., 1994, and Tao et al., 1994 are based on solving the optical properties of a multilayer optical system, in which experimentally obtained data for scattering are used as input parameters. In this approach numerical techniques are applied to calculate the absorption profile in the solar cell. In the model of Leblanc et al., 1994 the specular reflection and transmission coefficients of the electromagnetic field are assumed to be proportional to the classical Fresnel coefficients, where the proportionality factor depends on the amount of total diffused light. Consequently, specular light coherence is kept and specular interference effects are taken into account. The phase coherence between diffuse light and incident light is assumed to be lost at the interface. Further, the model can take into account scattering of the light for multiple passes. The model assumes two rough interfaces, the TCO/p^+(at the front) and the n^+/back reflector (at the back) interface, and symmetrical scattering coefficients between two adjacent layers. The effects of the texture in the whole stack are modeled by four parameters: the diffuse reflectances and the diffuse transmittances at the rough interfaces. These parameters are either experimentally measured at four visible wavelengths or adjusted to fit the experimental characteristics. The authors showed that the optical losses in the electrodes, in the doped layers and due to reflection, could be accurately calculated.

Tao et al., 1994 treats the a-Si:H solar cell on a textured substrate as a multi-thin-film structure. Since the thickness of the intrinsic a-Si:H layer is of the same order of magnitude as the grain size of the textured substrate and the doped layers are even much thinner, all the interfaces are rough and could, in principle, scatter the light. In order to calculate the optical properties of such a structure the multi-rough-interface optical model has been developed. The power of this model is that the it is easy to define wavelength dependent scattering properties of each rough interface of the system. Since in the solar cell the light can arrive from either side of a rough interface, four processes can take place: reflections at both sides of the interface and transmission of the light through the interface in both directions. These four processes are taken into account in the model and can be defined separately. Therefore this model is an excellent tool to investigate the effect of separate scattering processes in a solar cell, as demonstrated in Section 8.3. The model does not take into account coherence of the light. In the next section the multi-rough-interface optical model is described in more detail.

7.5.3 Multi-rough-interface optical model

The multi-rough-interface model is an extension of the multilayer system with flat interfaces that is treated in section 7.4.1. A rough interface is regarded as a flat interface with small disturbances that cause scattering. First we show how the reflectance and transmittance of the rough interface is modeled, followed by a single layer with two rough interfaces, and finally the complete multilayer structure with rough interfaces is described.

Single rough interface. In general a rough interface is represented by its reflectance and transmittance

$$R = f\left(\lambda, \tilde{n}_1, \tilde{n}_2, \Omega_{in}, \Omega_{out}, \sigma_r, m\right) \tag{7.48a}$$

$$T = f\left(\lambda, \tilde{n}_1, \tilde{n}_2, \Omega_{in}, \Omega_{out}, \sigma_r, m\right) \tag{7.48b}$$

which are functions of the wavelength λ, the complex refractive indices ($\tilde{n}_0$, $\tilde{n}_1$), the angle of incoming and outgoing light (Ω_{in}, Ω_{out}), the *rms* roughness σ_r, and the *rms* slope m.

A rough interface can scatter light in all possible directions and therefore in a system with rough interfaces the incoming and outgoing light from the rough interface is distributed in a half-sphere, as shown in Fig. 7.6. In the multi-rough-interface model the half-sphere can be divided in units that are described by a solid angle $\Omega = (\theta, \varphi)$ that ranges from 0 to 2π ($0 \leq \theta < \pi/2$, $-\pi \leq \varphi \leq \pi$). Under the assumption that the texture of the interface is random and that a sample is large enough to justifiably neglect effects of the edges, the light is expected to be distributed uniformly over all scattering angles and thus independent of φ. The light intensity F, which in general is a function of the solid angle, becomes a function of φ only,

$$F(\Omega) = F(2\pi \sin\theta) \tag{7.49}$$

This allows us to divide the half-sphere in Fig. 7.6 into N equal sub-cones with equal solid angle ($2\pi/N$). The solid angle distribution of F can be represented

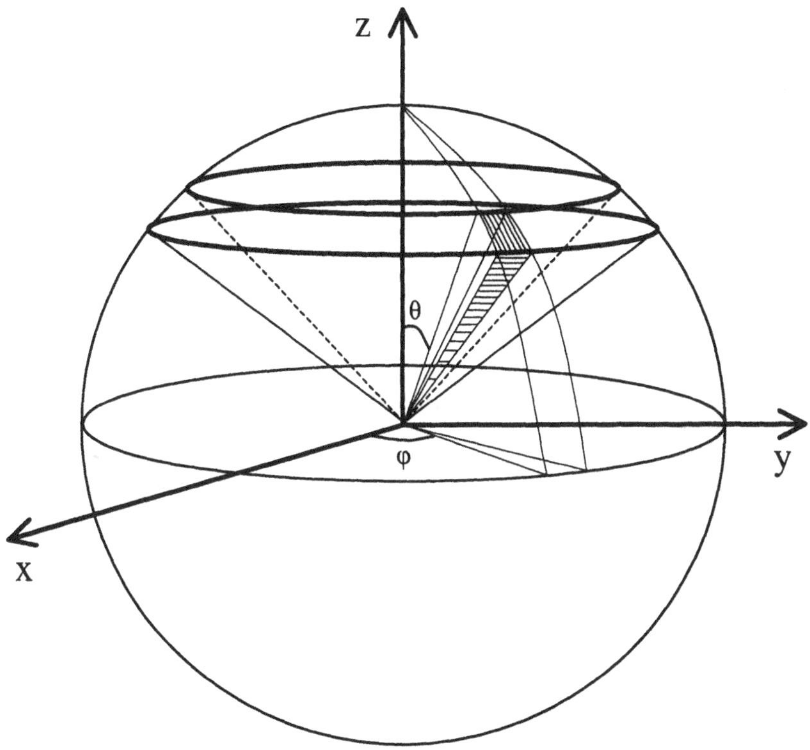

Figure 7.6. Solid angle distribution of incoming and outgoing light from a rough interface.

by an array of N elements, each element corresponding to a light intensity in a sub-cone. In order to take the scattering effect into account the reflectance and transmittance of the rough interface are represented by matrices of $N \times N$ elements, which include both specular and diffused parts of light.

The light can arrive at a rough interface from both media 1 and 2 (see Fig. 7.7), and therefore four processes are taken into account: the reflection in medium 1 (R_{11}), the reflection in medium 2 (R_{22}), the transmission from medium 1 to 2 (T_{12}), and the transmission from medium 2 to 1 (T_{21}). The coefficients R_{ii} and T_{ij} (i, j=1,2) are the reflectances and transmittances, respectively, that characterize the four processes. These can be represented by matrices of $N \times$ N elements. Energy conservation requires that

$$\sum_{\theta_{out}} \left[R_{11}\left(\theta_{in}, \theta_{out}\right) + T_{12}\left(\theta_{in}, \theta_{out}\right)\right] = 1 \quad \text{for every } \theta_{in} \tag{7.50a}$$

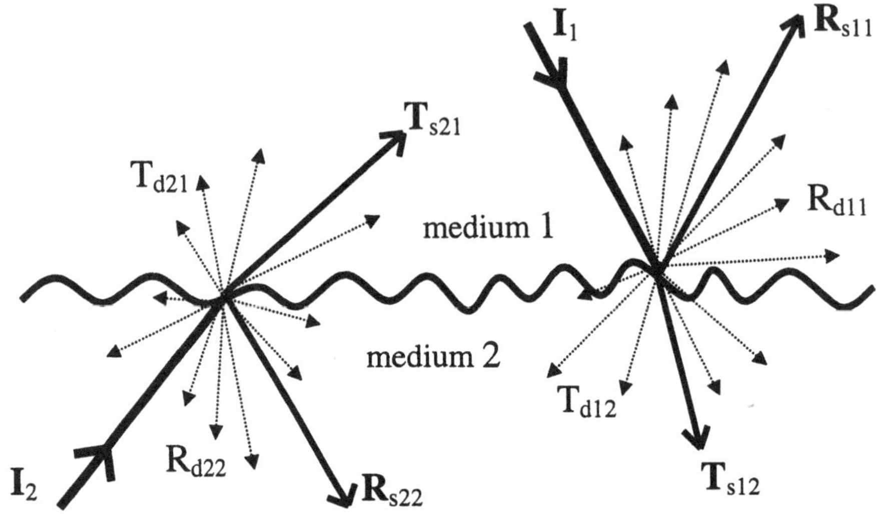

Figure 7.7. The reflection and transmission processes at a rough interface when light incidents on a rough interface from both sides.

$$\sum_{\theta_{out}} \left[R_{22}\left(\theta_{in},\theta_{out}\right)+T_{21}\left(\theta_{in},\theta_{out}\right)\right]=1 \quad \text{for every } \theta_{in} \tag{7.50b}$$

If we denote the up- and down-going light intensity in the first medium as F_1^- and F_1^+ and in the second medium as F_2^- and F_2^+ (see Fig. 7.8), the following relations can be formulated

$$F_1^- = R_{11}F_1^+ + T_{21}F_2^- \tag{7.51a}$$

$$F_2^+ = T_{12}F_1^+ + R_{22}F_2^- \tag{7.51b}$$

Scattering parameters of the rough interface. The following assumptions are used in the multi-rough-interface model in order to take scattering at rough interfaces into account:

1. A rough interface reflects the same amount of light as a flat interface. The

total reflected light (R_{tot}) at the rough interface, which is the sum of the specular and diffuse fractions of the reflected light, is equal to the light reflected by a flat interface (R_{flat}).

$$R_s + \sum_{\theta_{out}} R_d = R_{tot} = R_{flat} \tag{7.52}$$

2. The relation between the diffused part of the reflected light and the total reflected light is described as

$$R_d(\lambda, \theta_{in}, \theta_{out}) = C_{\mathrm{R}}(\lambda) f_{\mathrm{R}}(\theta_{in}) f_{\mathrm{R}}(\theta_{out}) R_{flat}(\lambda) \tag{7.53}$$

where C_{R} is the scattering coefficient that describes the ratio between the diffused part and the total of the reflected light.

A similar approach is used for transmitted light.

$$T_s + \sum_{\theta_{out}} T_d = T_{flat} \tag{7.54}$$

$$T_d(\lambda, \theta_{in}, \theta_{out}) = C_{\mathrm{T}}(\lambda) f_{\mathrm{T}}(\theta_{in}) f_{\mathrm{T}}(\theta_{out}) T_{flat}(\lambda) \tag{7.55}$$

where T_{flat} is the transmittance of a flat interface, C_{T} is the scattering coefficient for transmitted light, $f_{\mathrm{R,T}}\,(\theta_{in})$ describes the angle dependence of the incident light, $f_{\mathrm{R,T}}\,(\theta_{out})$ gives the dependence of the diffused light on the outgoing angle.

When the optical constants of two media are known, the reflectance R_{flat} and transmittance T_{flat} of the flat interface can be easily calculated using Eqns. 7.31 and 7.32. When knowing R_{flat} and T_{flat} and a set of scattering parameters $C_{\mathrm{R,T}}$, $f_{\mathrm{R,T}}\,(\theta_{in})$, $f_{\mathrm{R,T}}\,(\theta_{out})$) for all four processes that can take place at rough interface, the parameters that characterize these processes (reflectances R_{11} and R_{22} and transmittances T_{12} and T_{21}) can be calculated.

However, the required scattering parameters of the rough interfaces within a-Si:H solar cells are not well known. Only for some of the interfaces in a-Si:H solar cells scattering data have been measured (Schade and Smith, 1985b, Leblanc et al., 1994, Tao et al., 1994). The optical models that can take scattering into account in combination with numerical techniques such as inverse modeling (Ouwerling, 1987) are very useful to extract the scattering parameters of all rough interfaces in a-Si:H solar cells from simple measurements such as spectral response or total reflectance (Leblanc et al., 1994, Van den Berg et al., 1998).

Single layer with two rough interfaces. The optical system with two rough interfaces that separate three media is schematically shown in Fig. 7.8.

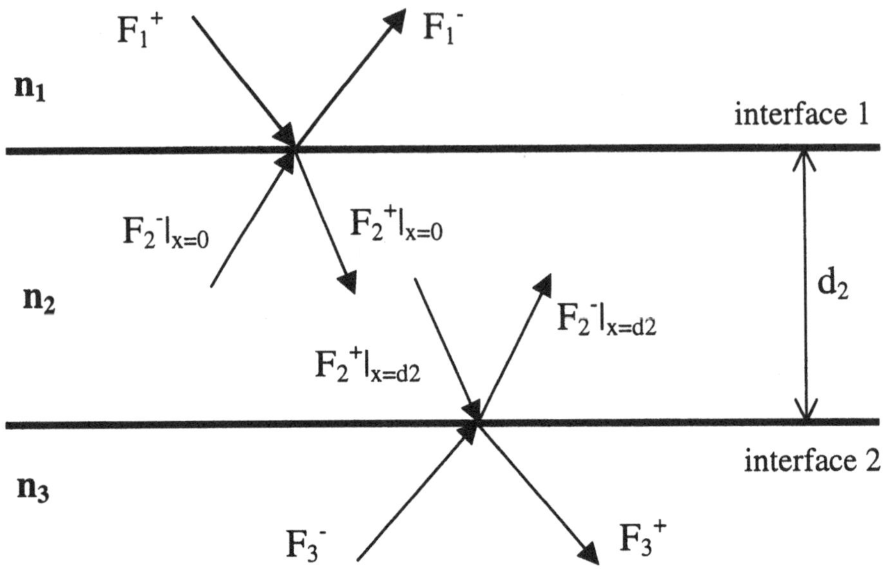

Figure 7.8. Incoming and outgoing light from a single layer with two rough interfaces.

The multi-rough-interface model assumes that the light is incoherent due to scattering and that both rough interfaces are described by Eqns. 7.51a and 7.51b. In addition, the absorption in the layer (medium 2) has to be taken into account

$$F_2^-\big|_{x=0} = M_2 F_2^-\big|_{x=d_2} \tag{7.56a}$$

$$F_2^+\big|_{x=d_2} = M_2 F_2^+\big|_{x=0} \tag{7.56b}$$

where M_2 is a diagonal matrix with elements of $\exp(-\alpha_2 d_2/\cos(\theta))$, in which α_2 is the absorption coefficient and d_2 is the thickness of the second medium.

The first interface is described by

$$F_1^- = R_{11,1}F_1^+ + T_{21}\, F_2^-\big|_{x_2=0} \tag{7.57a}$$

$$F_2^+\big|_{x_2=0} = T_{12}F_1^+ + R_{22,1}\, F_2^-\big|_{x_2=0} \tag{7.57b}$$

and the second interface by

$$F_2^-\big|_{x_2=d_2} = R_{22,2}\, F_2^+\big|_{x_2=d_2} + T_{32}F_3^- \tag{7.58a}$$

$$F_3^+ = R_{33,2}F_3^- + T_{23}\, F_2^+\big|_{x_2=d_2} \tag{7.58b}$$

The third subscript in R indicates the interface number. Assuming that $F_3^- = 0$, which means that there is no light incident from the third medium on the second medium, the effective R_{11}^* and T_{13}^* for the whole system can be easily found from Eqns. 7.56 to 7.58

$$R_{11}^* = R_{11,1} + T_{21} * M_2 * R_{22,2} * M_2 * X * T_{12} \tag{7.59}$$

$$T_{13}^* = T_{23} * M_2 * X * T_{12} \tag{7.60}$$

$$X = [E - R_{22,1} * M_2 * R_{22,2} * M_2]^{-1} \tag{7.61}$$

where E is the unit matrix. Similarly, when assuming that $F_1^+ = 0$, we obtain

$$R_{33}^* = R_{33,2} + T_{23} * M_2 * R_{22,1} * M_2 * \Xi * T_{32} \tag{7.62}$$

$$T_{31}^* = T_{21} * M_2 * \Xi * T_{32} \tag{7.63}$$

$$\Xi = [E - R_{22,2} * M_2 * R_{22,1} * M_2]^{-1} \tag{7.64}$$

For the general case where $F_1^+ = 0$ and $F_3^- = 0$, one can write

$$F_1^- = R_{11}^*F_1^+ + T_{31}^*F_3^- \tag{7.65a}$$

$$F_3^+ = T_{13}^*F_1^+ + R_{33}^*F_3^- \tag{7.65b}$$

Eqns. 7.65a and 7.65b, which describe the system with a single layer, have the form of Eqns. 7.51a and 7.51b, which describe a single interface. This implies that a single layer can be considered as a single interface which is characterized by an effective R and T.

Multilayer structure with rough interfaces. In the previous section it was demonstrated that an incoherent single layer can be considered as one effective interface. In the case of multilayer structure we iteratively apply this procedure of describing a layer as an effective interface to all subsequent layers of the system. In this way the light intensity F is determined at every interface of a multilayer structure (with m media), provided that the incoming light intensities F_1^+ and F_m^-, the optical properties of all media and the scattering properties of all interfaces are known. The absorption profile in the j-th medium for wavelength λ is

$$A_j(\lambda, x_j) = \int \frac{\alpha_j}{\cos(\theta)} F_{j,x_j}^- (\Omega)\, d\Omega + \int \frac{\alpha_j}{\cos(\theta)} F_{j,x_j}^+ (\Omega)\, d\Omega \tag{7.66}$$

Here, F_{j,x_j}^+ and F_{j,x_j}^- are the down- and up-going monochromatic light intensity at the depth x_j of the j-th medium. The normalized absorption profile is

$$A'_j(\lambda, x_j) = \frac{A_j(\lambda, x_j)}{\sum_\Omega F_1^+(\lambda, \Omega)} \tag{7.67}$$

The generation rate profile in the j-th layer for a certain light spectrum can be calculated by integrating the absorption profile over the spectrum

$$G_j(x) = \int A'_j(\lambda, x_j)\, \eta_g(\lambda)\, \Phi^0(\lambda)\, d\lambda \tag{7.68}$$

where $\Phi^0(\lambda)$ is the photon flux spectrum (for example AM1.5).

Fig. 7.9 shows the generation rate profiles in the reference solar cell (as in Fig. 8.1) for the case with only flat interfaces and the case with rough interfaces. The scattering coefficients used in the calculation are described in Section 8.3.

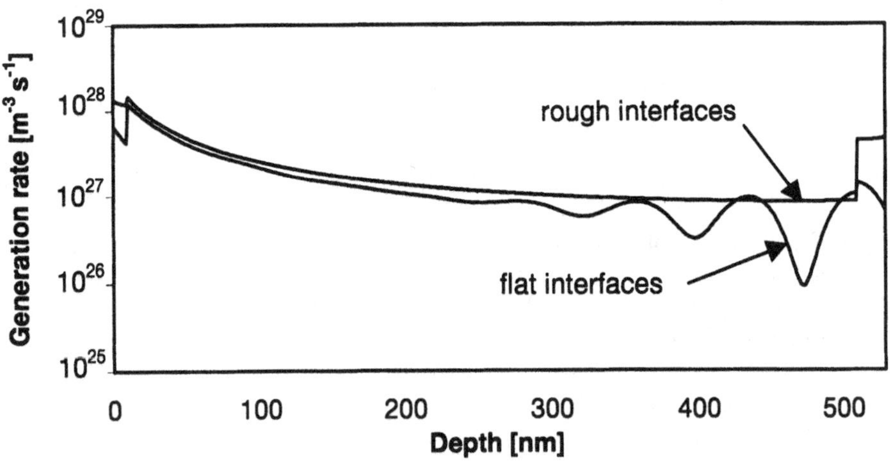

Figure 7.9. Generation rate profiles in the reference solar cell (as in Fig. 8.1) for the case with only flat interfaces and the case with rough interfaces.

References

Bennett, H.E., and J.O. Porteus, *Relation between surface roughness and specular reflectance at normal incidence*, J. Opt. Soc. Am. A **51** (1961) 1234.

Bennett, J.M., and H.E. Bennett, *Polarization*, in Handbook of Optics, Eds. W.G. Driscoll and W. Vaughan, Chap. 10, (Mc Graw-Hill, 1978).

Block, M., *Modellierung von Dünnschichtsolarzellen aus amorphem Silizium*, Ph.D. thesis (Fachbereich Physik der Phillips-Universität Marburg, Germany, 1993).

Bruns, J., *Die Entwicklung eines numerischen Simulationmodells für a-Si:H Solarzellen und seine Anwendung zur Analyse experimentell ermittelter Spektralcharakteristiken*, Ph.D. thesis (Fachbereich Elektrotechnik der Technischen Universität Berlin, Germany, 1993).

Chatterjee, P., and N. Palit, *Numerical optimisation of double junction solar cells with a-Si:H absorber layers*, Proc. of the 14th E.C. Photovoltaic Solar Energy Conference, 30 June - 4 July 1997, Barcelona, Spain, Eds. H.A. Ossenbrink, P. Helm, and H.A. Ehmann, (H.S. Stephens and associates, 1997) 517-520.

Curtins, H., and M. Favre, *Surface and bulk states determined by phototermal deflection spectroscopy*, in Advances in disordered Semiconductors, Vol. A (Amorphous Silicon an Related Materials), Ed. H. Fritzsche, (World Scientific, 1989).

Davies, H., *The reflection of electromagnetic waves from a rough interface*, Proc. IEEE part IV **101** (1954) 209-241.

Deckman, H.W., C.R. Wronski, H. Witzke, and E. Yablonovitch, *Optically enhanced amorphous silicon solar cells*, Appl. Phys. Lett. **42** (1983) 968-970.

Elson, J.M., H.E. Bennett, and J.M. Bennett, *Scattering from Optical Surfaces*, in Applied Optics and Optical Engineering, Vol. VII, Chap. 7, (Academic Press, 1979) 191-244.

Furlan, J.A., S. Amon, P. Popovič, F. Smole, *Internal opto-electrical properties of p-i-n a-Si:H solar cells on grooved TCO texture*, 1st World Conference on Photovoltaic Energy Conversion, (Proc. 24th IEEE PV Specialists Conference, Waikoloa, HI, USA, December 1994) 658-661.

Gordon, R.G., J. Proscia, F.B. Ellis, Jr. and A.E. Delahoy, *Textured tin oxide films produced by atmospheric pressure chemical vapor deposition from tetramethyltin and their usefulness in producing light trapping in thin film amorphous silicon solar cells*, Solar Energy Materials **18** (1989) 263-281.

Heavens, O.S., *Optical Properties of Thin Solid Films*, (Butterworths, London, 1995).

Hegedus, S., and W. Buchanan, *Effect of textured tin oxide and zinc oxide substrates on the current generation in amorphous silicon solar cells*, Proc. 25th IEEE PVSC, Washington D.C., May 13-17 (1996) 1129-1132.

Hulstrom, R., R. Bird and C. Riordan, *Spectral solar irradiance data sets for selected terrestrial conditions*, Solar Cells **15** (1985) 365-391.

Iida, H., N. Shiba, T. Mishuku, H. Karasawa, A. Ito, M. Yamanaka, and Y. Hayashi, *Efficiency of the a-Si:H solar cell and grain size of SnO_2 transparent conductive film*, IEEE Electron Dev. Lett. **EDL-4** (1983) 157-159.

Iida, H., T. Mishuku, A. Ito, and Y. Hayashi, *The structure of natively textured SnO_2 film and its application to an optical confinement-type a-Si:H solar cell*, IEEE Trans. Electron Dev. **ED-34** (1987) 271-276.

Leblanc, F., J. Perrin, and J. Schmitt, *Numerical modeling of the optical properties of hydrogenated amorphous-silicon-based p-i-n solar cells deposited on rough transparent conducting oxide substrates*, J. Appl. Phys. **75** (1994) 1074-1086.

Ley, L., *Photoemission and optical properties*, in: The Physics of Hydrogenated Amorphous Silicon, Eds. J.D. Joannopoulos and G. Lucovsky, Springer-Verlag, 1984.

Milonni, P.W., and J.H. Eberly, *Lasers*, John Wiley & Sons (1988).

Mizuhashi, M., Y. Gotoh and K. Adachi, *Texture morphology of SnO_2:F films and cell reflectance*, Jpn. J. Appl. Phys. **27** (1988) 2053-2061.

Morris, J., R.R. Arya, J.G. O'Dowd, and S. Wiedeman, *Absorption enhancement in hydrogenated amorphous silicon-based solar cells*, J. Appl. Phys. **67** (1990) 1079-1087.

Ouwerling, G.J.L., *The Profile/Prof2d User's Manual*, Delft University of Technology, 1987.

Porteus, J.O., *Relation between the height distribution of a rough surface and reflectance at normal incidence*, J. Opt. Soc. Am. **53** (1963) 1394-1402.

Sawada, T., N. Terada, T. Takahama, H. Tarui, M. Tanaka, S. Tsuda and S. Nakano, *Numerical approach for high-efficiency a-Si solar cells*, Solar Energy Materials and Solar Cells **34** (1994) 367-372.

Schade, H., and Z.E. Smith, *Mie scattering and rough interfaces*, Appl. Optics **24** (1985) 3221-3226.

Schade, H., and Z.E. Smith, *Optical properties and quantum efficiency of a-$Si_{1-x}C_x$:H/a-Si:H solar cells*, J. Appl. Phys. **57** (1985) 568.

Stiebig, H., A. Kreisel, K. Winz, N. Schultz, C. Beneking, Th. Eickhoff and H. Wagner, *Spectral response modelling of a-Si:H solar cells using accurate light absorption profiles*, 1st World Conference on Photovoltaic Energy Conversion, (Proc. 24th IEEE PV Specialists Conference, Waikoloa, HI, USA, December 1994) 603-606.

Tao, G., M. Zeman and J.W.Metselaar, *Accurate generation rate profiles in a-Si:H solar cells with textured TCO substrates*, Solar Energy Materials and Solar Cells 34 (1994) 359-366.

Van den Berg, J.H., M. Zeman and J.W. Metselaar, *Optical properties of a-SiGe:H solar cells on textured substrates*, J. Non-Cryst. Solids **227-230** (1998) 1262-1266.

Van den Heuvel, J., *Optical properties and transport properties of hydrogenated amorphous silicon*, Ph.D. Thesis (Delft University of Technology, The Netherlands, 1989).

Wallinga, J., *Textured transparent electrodes and series integration for amorphous silicon solar cells*, Ph.D. Thesis (Utrecht University, The Netherlands, 1998).

Wenham, S.R., and M.A. Green, *Silicon Solar Cells*, Progress in Photovoltaics: Research and Applications **4** (1996) 3-33.

Yablonovitch, E., and G.D. Cody, *Intensity enhancement in textured optical sheets for solar cells*, IEEE Trans. Electron Dev. **ED-29** (1982) 300-305.

8 INTEGRATED OPTICAL AND ELECTRICAL MODELING

Our analysis demonstrates that the physical mechanisms governing the operation of homojunction amorphous silicon p-i-n solar cells are more complicated than was originally thought.

—M. Hack and M. Shur, 1985

8.1 CALIBRATION OF MODEL PARAMETERS

In Chapters 6 and 7 the models that form a general mathematical description (also the term numerical model is used) of amorphous silicon material properties and physical processes in a-Si:H based devices were discussed. A computer program solves the model equations in order to obtain the desired information about device performance or properties. The usefulness of the simulation results and their predictive power strongly depend on the reliability of the input parameters that are required by the numerical model. Assigning the proper values to the input parameters is an important step in device modeling and is referred to as *calibration of model parameters*. The numerical model is calibrated well when a good agreement is reached between simulated and measured layer or device characteristics. A well calibrated computer model reproduces a broad range of experimental results.

One of the main difficulties in a-Si:H device modeling is that it requires a large number of parameters that describe the properties of materials in great detail. A typical a-Si:H device simulator that works with the DOS model for the R-G statistics, needs about thirty material parameters to define the properties of a-Si:H based layers. For the simplest single junction a-Si:H solar cell, which comprises three layers, a minimum of ninety input material parameters is thus needed. In addition, one has to define the device structure and the boundary conditions, so that the complete set of input parameters for the simulation of a single junction a-Si:H solar cell consists of about 100 parameters. The material parameters are in most cases dependent on the position in the device and/or a function of the energy level in the band gap. The fact that most of the input parameters are still not known precisely represents the main limitation to accurate modeling of a-Si:H solar cells. This parameter set does not yet include the optical properties of layers that were treated in Chapter 7. In addition to this set the optical generation profile, which is the result of optical modeling, is considered as an input parameter. For accurate computer simulation of solar cells, integrated optical and electrical modeling is required.

An additional difficulty is introduced by the inherent nature of hydrogenated amorphous silicon that its properties may vary from one laboratory to another and therefore it cannot be described by the same set of input parameters. For this reason there are no commonly accepted default values for a-Si:H based layers. The simulation results published by different groups are difficult to compare because different sets of input material parameters are used. The values of the parameters are based upon laboratory-dependent measurements complemented with data from literature. As mentioned in Chapter 3 a minimum set of characteristic properties has to be determined to qualify the material properties. Since most of these characteristic properties can be simulated, they represent an excellent starting point for the calibration of the input model parameters that describe properties of individual layers (Zeman et al., 1998).

It should be mentioned that the accuracy of the experimentally determined model parameters depends strongly on the type of experiment, i.e., an experiment can be very sensitive to a particular parameter while another experiment is not affected by this parameter at all. The relevance of a model parameter therefore strongly depends on the device structure used in a specific type of experiment. This feature is in fact widely used for the determination of model parameters. Modeling is first applied to investigate the sensitivity of model parameters on the result of a specific measurement. Based on the simulation analysis a test structure is designed such that the measurement is controlled by a particular model parameter. The designed structure is fabricated so that an accurate value of the model parameter can be extracted from the measured data. For example, this procedure was used by Zhu and Fonash, 1998 to extract the band offset between intrinsic a-Si:H and p^+ or n^+ microcrystalline silicon.

8.1.1 Calibration procedure of model parameters

A typical a-Si:H device simulator requires about 100 input parameters for calculating the characteristics of a simple single junction a-Si:H solar cell. Such a solar cell comprises three layers; p-type a-SiC:H, intrinsic a-Si:H, and n-type a-Si:H. No buffer layers or graded layers are taken into account since additional parameters are needed to describe them. A set of 100 parameters is a large set and it is extremely difficult to calibrate this set of parameters merely by fitting numerical simulations to measured characteristics of a solar cell. Using a simple analytical approach it was demonstrated that with different sets of input parameters one can obtain exactly the same dark and illuminated J-V characteristics of a solar cell (Willemen, 1998). Since most of the input parameters describe the opto-electronic properties of individual layers it is reasonable to use the experimental data of such individual layers as a *starting point* for calibration of these input parameters. It is assumed that the layers used in the solar cell do not deviate much from individual layers made under the same deposition conditions. The calibration procedure for the determination of the model input parameters based on measured properties of individual layers consists of the following steps (Zeman et al., 1998):
1. deposition of individual layers which comprise a reference solar cell
2. measurements of the material properties
3. fitting the model parameters to match the measured properties
4. simulation of the reference solar cell and other test devices and comparison with experimental results

The power of the procedure presented above is that it combines computer modeling with the results of simple, standard measurements of a-Si:H for extracting particularly those model parameters that are difficult to measure directly. The dark conductivity, the photoconductivity measured under standard AM1.5 illumination, the activation energy of the dark conductivity, and the absorption coefficient are the properties that are used in the fitting procedure. It is noted that high reliability experiments are required since the calibration procedure relies on the measured data. The process of fitting simulated to measured properties is accomplished completely by computer modeling: the calculations of measured properties, that are carried out by the device simulator, are combined with the parameter extraction technique (Zeman et al., 1994). This technique is called *inverse modeling*. In this technique, a selected set of input

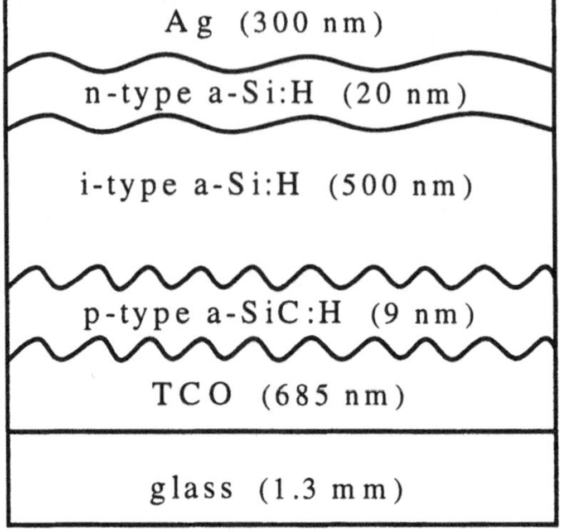

Figure 8.1. Schematic structure of the a-Si:H p-i-n reference cell.

parameters, called *fitting parameters*, is adjusted by a non-linear optimization driver in order to obtain an optimum match between the simulated and the measured data. This technique is described in section 8.1.2.

The input parameters that are difficult to measure directly are those that describe the DOS distribution in the band gap of a-Si:H based materials. Modeling of the subbandgap absorption coefficient in combination with the electrical properties is found to be sensitive to these model parameters and therefore suitable for extracting their values. Using a single set of input parameters, which includes the extracted values of the DOS distribution characteristics as the fitting parameters, a good agreement between the measured and simulated properties of individual a-Si:H layers was obtained (Zeman et al., 1998).

The sets of input parameters describing the individual layers was combined to create a set of input parameters for simulating a reference solar cell. Schottky barriers were introduced at the front and back contacts of the cell as the boundary conditions. The front contact barrier is situated at the TCO/p-layer interface while the back contact barrier is located at the interface of the n-layer and the metal back contact. With this set of input parameters, presented in Table 8.1, a good agreement between the measured and simulated J-V characteristics was reached. The structure of the reference cell is schematically depicted in Fig. 8.1 and the measured and simulated AM1.5 J-V characteristics are shown in Fig. 8.2. The solar cell presented in Fig. 8.1 is referred to as

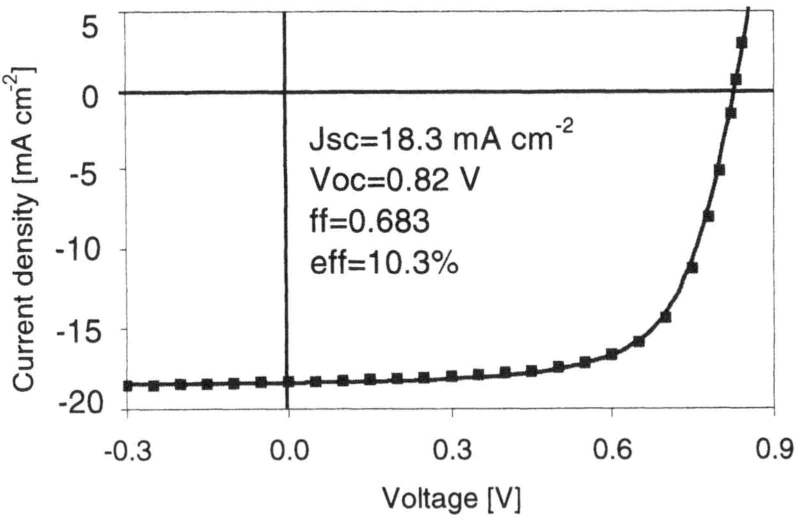

Figure 8.2. The measured (dots) and simulated (continuous curve) illuminated J-V characteristics of the a-Si:H p-i-n reference cell.

the *reference solar cell* throughout this book.

8.1.2 Extraction of model parameters by inverse modeling

Inverse modeling is an effective and powerful way to extract values of model input parameters that cannot be measured directly or accurately. Usually a computer program (or a device simulator) calculates device output characteristics from given input parameters. This mode is called forward modeling and the device simulator then is a forward model. In the inverse modeling mode the results obtained from forward modeling combined with measured data, are input and the values of the fitting parameters are obtained as output. In this simulation scheme, in addition to a forward model, another computer model is used, a so called optimization driver. The non-linear optimization driver minimizes the error between the simulated and measured characteristics in an iteration procedure by adapting the values of the fitting parameters. An example of the non-linear optimization driver is PROFILE that was developed at Delft University of Technology (Ouwerling, 1987). The algorithm of the Modified Damped Least Square (MDLS) optimization method (Meyer and Roth, 1972) is implemented in PROFILE. Besides the extracted values of selected fitting parameters PROFILE has several statistical tools that provide additional information about sensitivity and mutual dependence of the fitting parameters. The fitting parameters form a parameter space in which the optimization method looks for the minimum of the least squares error between the measured and simulated data. It should be mentioned that several minima may be present in the parameter space and therefore no unique solution or set of extracted values can be obtained.

Zeman et al., 1994 applied the inverse modeling technique to determine a set of input parameters that describe the single junction a-Si:H solar cells. The values of selected fitting parameters were obtained from fits to several J-V curves simultaneously. They found that the most sensitive input parameters from the group of selected fitting parameters was the DOS at the mobility edges and the density of defect states in the intrinsic layer.

8.1.3 Sensitivity study of a-Si:H solar cell model parameters

The combination of a forward model with the inverse modeling technique is a very effective way for extracting values of input parameters for the forward model provided that the proper set of fitting parameters is selected. The selection of the fitting parameters depends on the type of experiment and the sensitivity of this experiment to the model parameters. The chosen parameters should be those to which a material property or solar cell characteristics are the most sensitive. For this reason it is necessary to identify the importance of the model parameters on the results of standard measurements of a-Si:H solar cells. In the literature several sensitivity studies for a-Si:H solar cells have been published: McElheny et al., 1991 have studied the sensitivity of the quantum efficiency of a-Si:H Schottky barrier solar cells and Chatterjee, 1994 has studied the sensitivity of the illuminated J-V characteristics of an a-Si:H p-i-n cell. In both papers the sensitivity of only a few solar cell characteristics to parameter changes were investigated.

A complete study of the sensitivity of a-Si:H single junction solar cell characteristics to the model parameters was carried out by Willemen et al., 1996. In this study the sensitivities of three different solar cell characteristics were analyzed: dark J-V, illuminated J-V, and external quantum efficiency, QE. For each of these characteristics the response parameters were defined. The diode ideality factor, n, the saturation current density, J_0, and the series resistance factor, F_s, are the response parameters for the dark J-V characteristics. The open circuit voltage, V_{oc}, the short circuit current density, J_{sc}, and the fill factor, FF, are the response parameters for the illuminated J-V characteristic, and the integrated QE between 300 nm and 550 nm wavelength and the integrated QE between 550 nm and 800 nm wavelength. The sensitivity of a response parameter to a model parameter is determined from the difference between the response of a baseline parameter set and the response obtained with a set where a particular model parameter is changed to its boundary value. Some of the results of this study are briefly discussed below Willemen, 1998.

The dark J-V response parameters n and J_0 are most sensitive to the i-layer dangling bond density and capture cross sections, therefore measurements of the dark J-V characteristics are suitable experiments to study the mid-gap density of states in the i-layers in a-Si:H pin diodes. A little less influential are the mobility gaps of the p-layer and i-layer. The series resistance factor F_s is most sensitive to the p-layer parameters that control the conductivity of this layer, such as the hole mobility and the activation energy.

The most important model parameters for the V_{OC} are the mobility gap of the i-layer and the mobility gap and the activation energy of the p-layer. The large influence of the mobility gap and the effective density of extended states in the i-layer is related to their impact on the maximum possible separation of the i-layer quasi-Fermi levels due to optical carrier generation. Since V_{OC} also depends on the carrier recombination in the device, a relatively strong influence of the band tail parameters, the defect density, and the capture-rate coefficients on the V_{OC} is observed.

The short circuit current density J_{sc} is sensitive to the optical parameters of all layers and to the irradiance of the illumination. These parameters determine the absorption profile in the cell and J_{sc} is proportional to the absorption in the intrinsic layer. For monochromatic illumination at 400 nm (blue light) the optical properties of the TCO and the p-layer and the electron mobility in the i-layer are influential parameters. In the case of illumination at 600 nm (red light), J_{sc} is sensitive to the optical parameters of the TCO and the i-layer, and the hole mobility in the i-layer. The stronger influence of the hole mobility in the case of red light illumination is due to the fact that the average distance, that the holes have to travel through the cell towards the collecting p-layer, is larger then in the case of the blue light illumination. Red light is almost uniformly absorbed in the intrinsic layer while blue light is mostly absorbed in the front part of the intrinsic layer. When the light enters the solar cell through the p-layer then the electron-hole pairs are generated in the vicinity of the p-layer and the average distance that the holes have to cover to reach the p-layer is much smaller. Since the hole mobility is less than the electron mobility in intrinsic a-Si:H it is desired that the majority of the holes is generated close to the collecting p-layer. This is the case when the light enters the solar cell through the p-layer. The simulations confirm the experimental observation that devices illuminated through the p-layer almost always have a higher conversion efficiency than when illuminated through the n-layer. The J_{sc} generated at 600 nm is the only response parameter on which the n-layer thickness has a significant influence. The n-layer acts as a parasitic absorber of long wavelength light (above 600 nm) that is transmitted through the i-layer.

The *FF* of the cells under standard AM1.5 illumination is sensitive to those parameters that determine the conductivity of the layers such as the activation energy of the p-layer, the effective density of states, and the mobilities of the carriers in the p-layer and the intrinsic layer. The parameters that control the strength of the electric field across the i-layer such as the tail states parameters and the concentration of dangling bonds show a strong influence on the *FF*. Also the barrier between the TCO and the p-layer has a large influence on the *FF*.

Another valuable result of the sensitivity study is the identification of those model parameters that do *not* show influence on the investigated response parameters. The parameters of the n-layer demonstrate negligible influence on the response parameters. This fact explains why little attention is paid to the n-layer in the literature. Also in experimental work the n-layer is found not to be as critical as the p-layer and/or the intrinsic layer.

The results of the sensitivity study of the model parameters show that the solar cell performance is mainly influenced by the properties of intrinsic layer and p-type layer. This fact is reflected in the large amount of modeling work that studies how these properties influence the overall performance of the solar cell. The effect of the p-layer thickness, the front contact barrier height, the p/i interface states was studied in detail by several groups (Tasaki et al., 1988, Arch et al., 1990, Chatterjee, 1994, Chatterjee, 1996, Smole and Furlan, 1992, Smole et al., 1994).

In summary, the benefits of the results of the sensitivity study are that
i) the fitting parameters in the model calibration procedure that are relevant to a particular measurement when using inverse modeling techniques can be properly chosen,
ii) the most influential model parameters that can be translated into the technological parameters are revealed,
iii) the knowledge of the sensitivity of various response parameters to the model parameters can be used to design device structures for accurate measurements and extraction of model parameter values.

8.1.4 Set of model input parameters

Before simulating any device, the materials that are used to build the structure as well as the contacts have to be described. The minimum set of input parameters that is needed to simulate a single junction a-Si:H solar cell without any buffer and graded layers is listed in Table 8.1. Most of the values of these parameters resulted from the calibration procedure described in Section 8.1.1.

Table 8.1. The set of input parameters for modeling the reference single junction a-Si:H solar cell. The symbols used in the table are explained in Section 6.4.

Material and contact parameters	*p-layer*	*i-layer*	*n-layer*
Thickness (nm)	9	500	20
Relative permittivity	7.2	11.9	11.9
Electron affinity (eV)	3.90	4.00	3.99
Mobility gap (eV)	1.95	1.78	1.80
Electron mobility ($10^{-4}m^2/Vs$)	20.0	20.0	20.0
Hole mobility ($10^{-4}m^2/Vs$)	5.0	5.0	5.0
Effective DOS in CB (m^{-3})	1.0×10^{26}	1.0×10^{26}	1.0×10^{26}
Effective DOS in VB (m^{-3})	1.0×10^{26}	1.0×10^{26}	1.0×10^{26}
Activation energy (eV)	0.43	0.83	0.20
Front contact Schottky barrier (eV)		1.24	
Back contact Schottky barrier (eV)		0.25	

Tail states parameters	*p-layer*	*i-layer*	*n-layer*
DOS at CB mobility edge (m^{-3}/eV)	2.0×10^{27}	8.0×10^{27}	1.0×10^{27}
DOS at VB mobility edge (m^{-3}/eV)	1.0×10^{27}	4.0×10^{27}	2.0×10^{27}
CB tail characteristic energy (eV)	0.180	0.032	0.070
VB tail characteristic energy (eV)	0.090	0.047	0.160
Capture coefficient neutral states (m^3/s)		7.0×10^{-16}	
Capture coefficient charged states (m^3/s)		7.0×10^{-16}	

Dangling bond states parameters	*p-layer*	*i-layer*	*n-layer*
Model		SDM (Gaussian distribution)	
Standard deviation	0.144	0.144	0.144
Total density (m^{-3})	8.0×10^{24}	5.0×10^{21}	2.0×10^{25}
Position of $E_C^{mob} - E_{DB}^{+/0}$ (eV)	-0.70	-0.89	-1.40
Correlation energy (eV)	0.20	0.20	0.20
Capture coefficient neutral states (m^3/s)		3.0×10^{-15}	
Capture coefficient charged states (m^3/s)		30.0×10^{-15}	

It is important to note that these values are not the commonly accepted default values for a-Si:H based layers but they can serve as a reference.

8.2 UNDERSTANDING a-Si:H SOLAR CELL PERFORMANCE

The standard measurements used to evaluate the performance of solar cells are the J-V measurements in the dark and under different illumination conditions. The solar cells are commonly characterized and compared by external parameters such as V_{oc}, J_{sc}, FF, and the conversion efficiency, which are determined from the J-V measurement under standard AM1.5 illumination. Additional measurements are the spectral response measurements that can be carried out under different bias voltage and bias illumination. All these measurements can be simulated by an amorphous silicon device simulator. In addition to these results, the device simulator can provide the user with extra information that is not available from experiments. This can be the electric field, the concentration of free charge carriers, the space charge or recombination rate as a function of the position in the solar cell. This information in particular helps to understand the processes that take place in the device and to establish a link between the properties of the layers expressed in the input model parameters and the overall device performance.

Hack and Shur, 1985 were the first to present a comprehensive computer simulation of amorphous silicon alloy p-i-n solar cells by analyzing the free carrier concentration, trapped carrier concentration, space-charge concentration, recombination rate, and electric field profiles as well as corresponding current-voltage characteristics for a variety of device parameters and illumination conditions. They showed that the principle loss of carrier collection is caused by bulk recombination, which is governed by the distribution of free carriers, and the performance is determined by the transport properties of the limiting carrier, i.e., holes in intrinsic material. Taking into account that the hole is the limiting carrier they also quantitatively demonstrated that solar cells illuminated through the p-layer have a higher conversion efficiency then when illuminated through the n-layer.

To demonstrate the power of computer modeling to provide additional information on the physics of the device we shall analyze the effect of the dangling bond density in the intrinsic layer on the performance of a p-i-n a-Si:H solar cell. This is also one of the main parameters that is used to simulate the effect of light soaking, which generally leads to degradation of a-Si:H solar cells. The results that are presented are obtained using the values of the input parameters listed in Table 8.1. In the case of the standard DOS model with a Gaussian distribution of defect states a constant value for the defect concentration throughout the intrinsic layer was used. In the simulations the concentration of defect states was varied from 5×10^{21} m^{-3} to 1×10^{23} m^{-3}. In the case of the Defect Pool Model the concentration of defect states was calculated as a function of the position of the Fermi level in the device. The defect density profiles in the intrinsic layer that were used in the simulations are shown in Fig. 8.3.

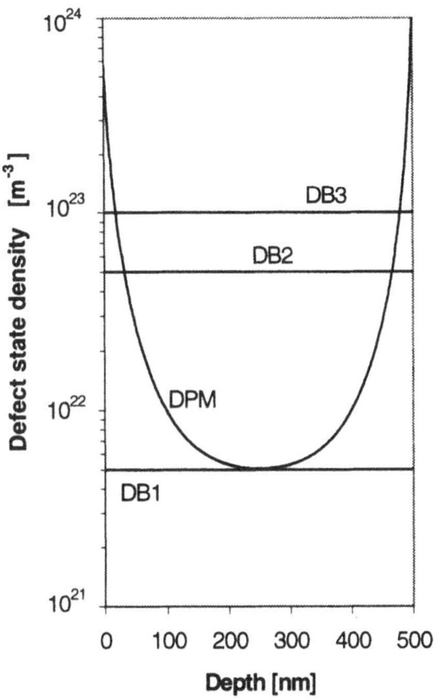

Figure 8.3. The defect density profiles in the intrinsic layer of a-Si:H p-i-n solar cell.

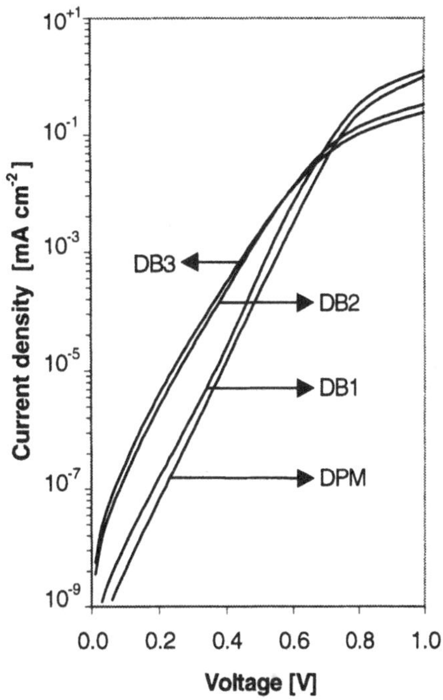

Figure 8.4. The dark J-V characteristics for different defect density profiles in the intrinsic layer of a-Si:H p-i-n solar cell.

The resulting J-V characteristics in the dark and under standard AM1.5 illumination that are measurable quantities are plotted in Fig. 8.4 and Fig. 8.5, respectively. The external parameters of solar cells were determined from the illuminated J-V characteristics and are summarized in Table 8.2. From Fig. 8.5 it is evident that an increase of the total density of defect states in the intrinsic layer deteriorates the performance of the cell. All external parameters of the cell are affected. This result clearly explains the continuous effort of researchers to produce a stable a-Si:H material with a low density of defect states.

Computer modeling allows us to extend the analysis of the cell beyond a mere evaluation of the overall performance. The operation of an a-Si:H solar cell is based on the drift of the photogenerated carriers in the electric field

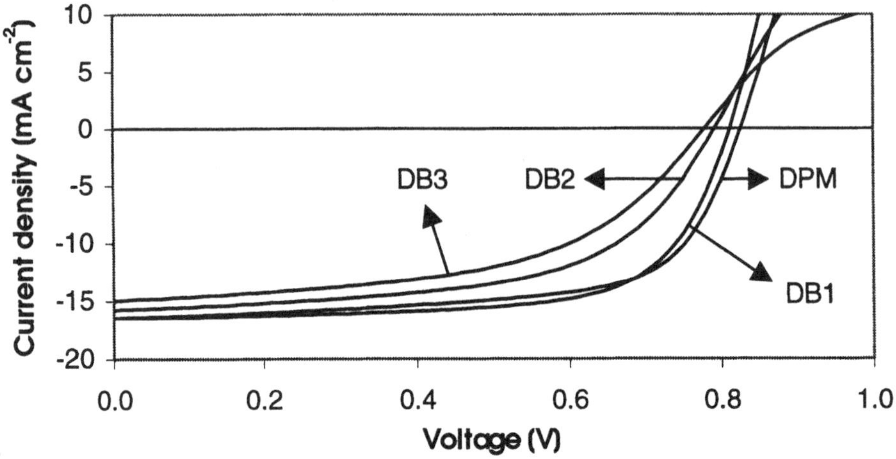

Figure 8.5. The J-V characteristics of a-Si:H p-i-n solar cell under AM1.5 illumination for different defect density profiles in the intrinsic layer.

across the intrinsic layer. The electric field profile in the intrinsic layer under short circuit current conditions is plotted for different defect states distributions in Fig. 8.6. An increased defect density leads to a reduced electric field in the bulk of the intrinsic layer, which results in poorer carrier collection from this part of solar cell. The poorer collection is associated with an increased recombination rate. This can be seen in Fig. 8.7 where the recombination rate profile in the intrinsic layer under short circuit current conditions is shown. The maxima in the recombination rate profiles roughly correspond to the minima in the electric field profiles. The recombination rate

Table 8.2. The simulated external parameters of p-i-n a-Si:H solar cell for different defect density profiles in the intrinsic layer.

Symbol	N_{db} [m^{-3}]	J_{sc} [mA/cm^{-2}]	V_{oc} [V]	J_{max} [mA/cm^{-2}]	V_{max} [V]	FF	eff. [%]
DB1	5.0×10^{21}	16.49	0.81	14.05	0.65	0.68	9.15
DB2	5.0×10^{22}	15.79	0.79	12.21	0.59	0.57	7.19
DB3	1.0×10^{23}	14.99	0.78	11.20	0.55	0.53	6.19
DPM		16.48	0.82	13.35	0.67	0.66	8.99

also depends on the total density of defect states and the capture coefficients.

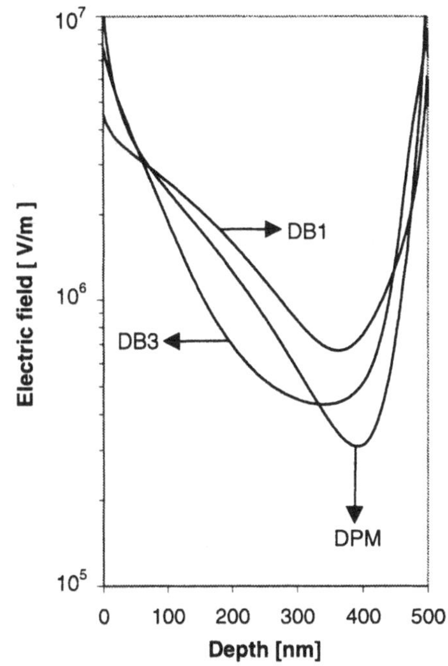

Figure 8.6. The electric field profile in the intrinsic layer of an a-Si:H p-i-n solar cell under short circuit current conditions.

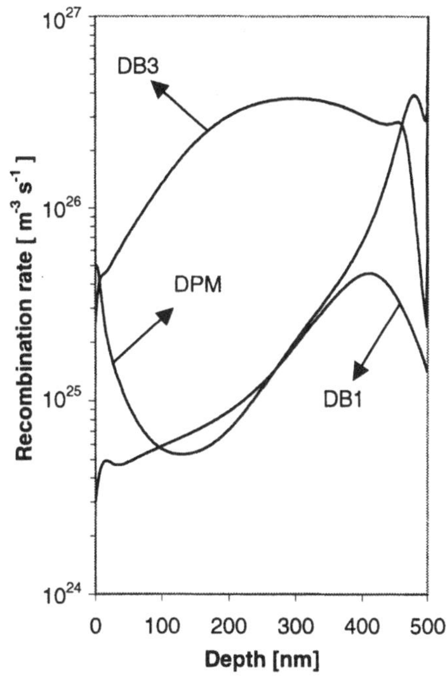

Figure 8.7. The recombination rate profile in the intrinsic layer of an a-Si:H p-i-n solar cell under short circuit current conditions.

It is interesting to note that when the cell is illuminated the lowest electric field and the highest recombination rate are positioned close to the i/n interface. This result suggests that one has to pay attention not only to the p/i interface but also to a proper design of the i/n region to improve the collection from this part of the solar cell.

Not only an increase in the total density of defect states influences the performance of the cell but also the spatial distribution of the defects plays an important role. This is evident from the comparison of the simulated quantities corresponding to DB1 and DPM parameters. Although both sets of parameters lead to similar external parameters of the solar cell, the internal quantities are quite different. The straightforward advantage of using the defect states profile calculated according to the Defect Pool Model is that one can avoid the need of placing an extra defect layer in the p/i region in order to account for the losses at this interface.

8.3 OPTIMIZATION OF a-Si:H SOLAR CELLS

8.3.1 Optical design of a-Si:H solar cells

Optical device modeling has become a powerful tool for gaining deeper knowledge about the optical properties and behavior of a-Si:H solar cells and for contributing to the optimal optical design of solar cell structures. The aim is to design a solar cell structure in which the absorption of incident light in the active part of the solar cell is maximized and the absorption in the non-active layers is minimized. The advantage of a modeling approach over an experimental approach is that the absorption of light can be determined seperately for all individual layers and that the influence of the optical parameters on the absorption profile in the cell can be evaluated separately for each layer. Modeling can provide important information for the optimal optical design that cannot be obtained experimentally.

One can use this approach if the optical model is well calibrated, i.e., if the input parameters of the model reproduce the available experimental data. As mentioned in the previous section, computer modeling in combination with simple opto-electronic measurements is a powerful tool for extracting those parameters that are at present difficult to obtain experimentally. The relatively simple measurements that convey information for the extraction of optical model parameters, are the total reflectance and the absolute external quantum efficiency.

The absolute external quantum efficiency is defined as the number of charge carriers collected (from all layers of device) per incident photon for a range of wavelengths. This quantum efficiency is defined as

$$QE_{ex,abs}(\lambda) = \sum_{layers} QE_{op}(\lambda)\eta_g(\lambda)QE_{el}(\lambda) \tag{8.1}$$

where QE_{op} is the optical quantum efficiency and is a measure for the probability for a photon to be absorbed. QE_{el} is the electrical quantum efficiency and reflects the probability for a photogenerated carrier to be collected. The term quantum efficiency (QE) used throughout this book refers to the absolute external quantum efficiency defined above.

Usually, by applying a sufficient negative bias voltage in the QE measurement, all photogenerated carriers are collected. Under these conditions, QE_{el} is unity and the measurement of $QE_{ex,abs}$ thus represents the optical quantum efficiency. Since we assume that only the photons that are absorbed in the active layer of solar cell contribute to the measured current, the optical quantum efficiency corresponds to the total absorptance of photons in the active layer. This can be calculated using purely optical modeling, which is less time consuming and requires fewer input parameters than the complete electrical modeling.

Fig. 8.8 demonstrates the capability of optical modeling to evaluate a solar cell structure as an optimally designed optical system. It shows the results of investigating the effects of the scattering model parameters on the QE of the reference cell and the losses in the non-active layers. The multi-rough-interface model that is described in this book (Section 7.5.3) was used to calculate absorption profiles in the reference solar cell. The structure of the reference solar cells is given in Fig. 8.1. The optical constants of the layers that were used in simulations are shown in Fig. 7.1. The absorption profile in the individual layers of the cell was integrated over the thickness of the layer in order to obtain the total absorption. The total absorption in the intrinsic a-Si:H layer calculated per incident photon is compared to the measured QE. The total absorption in the other layers, which is considered optical loss, is in this figure expressed in units of QE.

Fig. 8.8a shows the case of a cell with flat interfaces only. The interference pattern is absent in this representation because the multi-rough-interface model works with fully incoherent propagation of light. Figs. 8.8b to 8.8d show several cases in which scattering has a major effect on the QE of the cell. These cases are: scattered reflection of light from n/metal interface (Fig. 8.8.b), scattered reflection from the TCO/p interface (Fig. 8.8.c), and scattered transmission at TCO/p interface (Fig. 8.8.d). In all three cases it is assumed that the interface acts as a perfect diffuser for the particular process, i.e., all photons are scattered and the angular distribution of the scattered photons is Lambertian. This is an example in which modeling enables us to investigate structures that are difficult to fabricate or that can not be fabricated at all in practice and thus to evaluate the effect of particular parameters separately.

The simulations demonstrate the trends in the optical behavior of the solar cell when scattering is introduced. The influence of scattering can best be illustrated by discussing two wavelength regions: the short wavelength region from 350 nm to 550 nm and the long wavelength region from 550 nm to 900 nm. In the short wavelength region, scattering at the TCO/p interface causes both an increase (scattered reflection) and decrease (scattered transmission) in the QE as demonstrated by Fig. 8.8c and Fig. 8.8d, respectively. The resulting effect of scattering at TCO/p interface depends on in which scattering process dominates. In the long wavelength region, scattering at the TCO/p interface always has a beneficial effect on the QE. Scattering at the n/Ag interface affects the QE in the long wavelength region. It increases the QE in this wavelength region but also causes increased absorption in the n-layer and in the metal. Fig. 8.8b and Fig. 8.8d demonstrate that upon applying scattering in the cell the major loss in the long wavelength region is absorption in the metal contact layer. Chatterjee and Palit, 1997 arrived to a similar conclusion. With the use of the optical model of Leblanc et al., 1994 they showed that for weakly absorbed red radiation, the major loss component is the absorption in the Al metal contact. The absorption in the TCO and doped layers occurs mainly in the short wavelength region and is enhanced by scattering at the TCO/p textured interface. Hishikawa et al., 1996, who used optical simulations of a solar cell with flat interfaces, presented a relation between the optical absorption in the transparent electrode (SnO_2) and the experimentally determined optical loss. They concluded that the absorption in the TCO layer, which is enhanced by the optical confinement effect of the textured surface, represents a major loss corresponding to > 5 mA/cm^2 in the short circuit current density. They proposed the following optical designs to improve J_{sc}: i) an increased reflectance at a-Si/SnO_2 interface (preferably total reflection), ii) an increase in average scattering angle at the front and rear surfaces, combined with a reduction in thickness and absorption coefficient of SnO_2. The results of simulations that take scattering into account reveal that the major loss component is absorption in the rear metal contact and not within the TCO layer. This means that attention has to be paid to optimization of the back contact.

In order to calculate accurate absorption profiles in a solar cell with textured interfaces the proper values of scattering coefficients of the rough interfaces have to be known. The following strategy is used to determine the values of scattering coefficients. The coefficients of diffuse reflectances of the rough interfaces are calculated from Eqn. 7.47. The TCO/p interface has a more pronounced roughness than the top surface of the deposited cell (see Fig. 4.5). The value of the *rms* roughness for the TCO/p and p/i interface was measured to be 30 nm, while for the i/n and n/Ag interface it was 10 nm. The calculated scattering coefficients for the diffuse reflectances from the textured interfaces as a function of wavelength are plotted in Fig. 8.9. The coefficients for diffuse transmittance were taken constant at 0.1 for all interfaces. Using these values of scattering coefficients a good agreement between the simulated and measured QE's of the reference solar cell was reached (see Fig. 8.10). To establish a relation between the interface roughness and the corresponding scattering parameters is one of the most important issues in the present research.

Modeling was used to optimize highly reflective TCO/metal back contacts for a-Si:H solar cells (Morris et al., 1990, Tao et al., 1992, Stiebig et al., 1994). Morris et al., 1990 found that the texture imparted to the metallic interface may result in a significant decrease in the reflectance of the rear contact. A thin

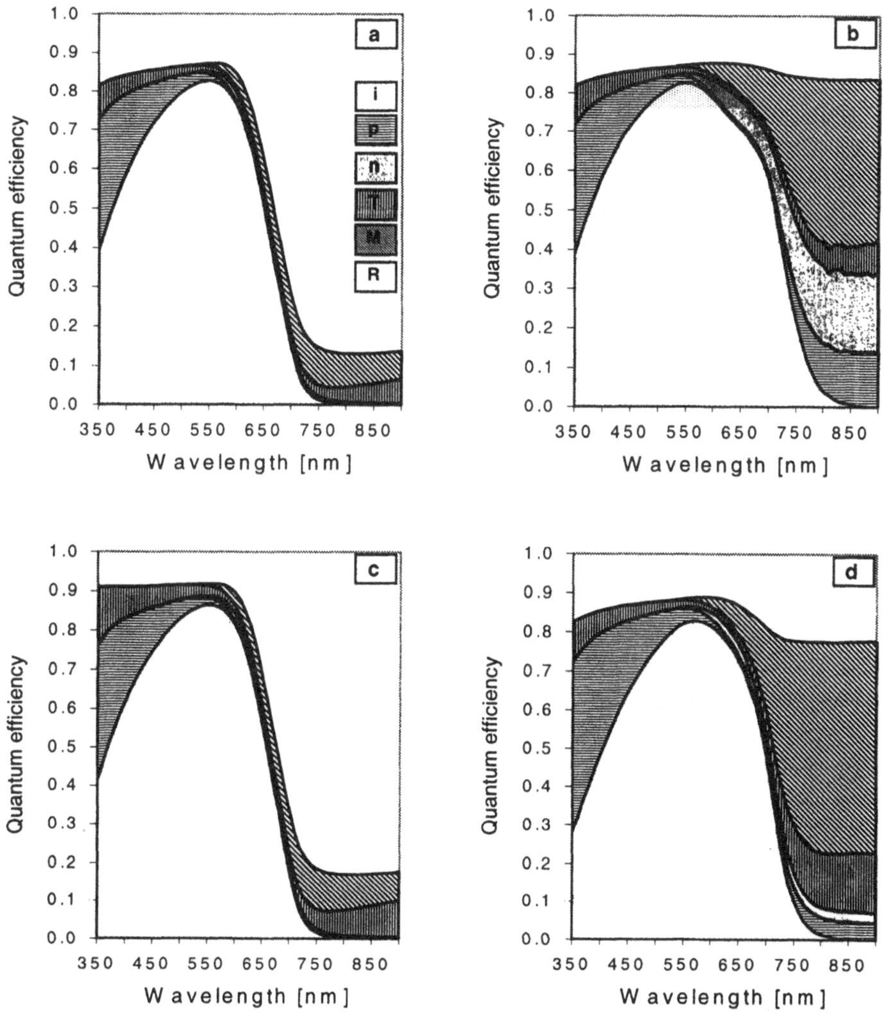

Figure 8.8. The total absorption in the separate layers and total reflection of the reference a-Si:H single junction cell expressed as quantum efficiency. Abbreviations: i is intrinsic a-Si:H layer, p is p-type a-SiC:H layer, n is n-type a-Si:H layer, T is SnO_2:F TCO layer, M is metal contact, and R is the total reflection of the cell. a) all interfaces in the cell are flat b) n/metal interface is rough and considered as a perfect diffuser c) TCO/p interface is rough and only the light reflected back to the TCO is scattered d) TCO/p interface is rough and only the light transmitted to the p-layer is scattered.

layer of indium-tin-oxide (ITO) inserted between the a-Si:H n^+ layer and the

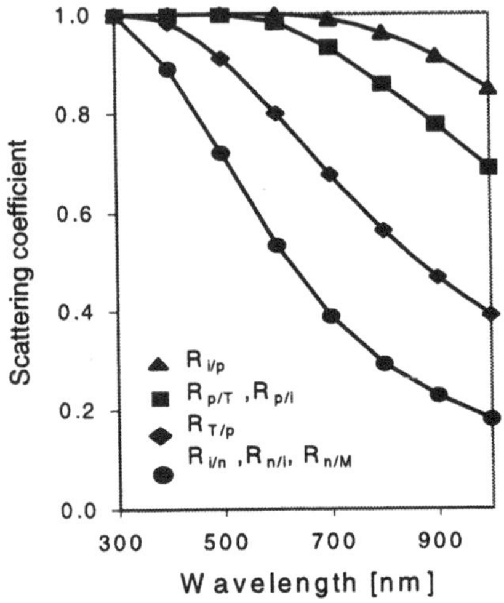

Figure 8.9 Scattering coefficients of reflected light for different textured interfaces in the reference solar cell.

contact metal results in critical trapping of scattered light and thus reduces the absorption of light in the rear contact. Tao et al., 1992 used the following approach to optimize the TCO/metal back contact for obtaining high reflectance of a-Si:H and back contact interface. First they calculated the reflectance of different interfaces such as a-Si:H/Al, a-Si:H/Ag, a-Si:H/TCO/Al, air/TCO/Al, air/a-Si:H/Al and air/a-Si:H/TCO/Al. The results show that the reflectance averaged over a wavelength range from 500 nm to 800 nm is less than 70 % for the a-Si:H/Al interface and about 85 % for the a-Si:H/Ag interface. The average reflectance exceeds 90 % when an ITO layer with a thickness within the range of 50 nm to 90 nm is inserted between a-Si:H and Al or Ag. In the next step they compared the calculated results with the measured values of the reflectance to verify the simulation results. In the third step the ITO thickness in the a-Si:H solar cell was optimized by modeling the complete cell. The optimum ITO thickness was found to be 70 nm for both a-Si:H/Al and a-Si:H/Ag interfaces. Finally, the optimized back contact was applied to the solar cell. The implementation of ITO/Al back contact in a solar cell on a flat substrate led to 11 % relative improvement in J_{sc}. The simulations of the reflectance as a function of the incident angle showed that the reflectance of the a-Si:H/ITO interface is almost unity for incident angles larger than 30 degrees. This implies that when applying TCO/metal back contacts on the rough a-Si:H solar cell, the TCO thickness and the type of metal are not critical for scattering the

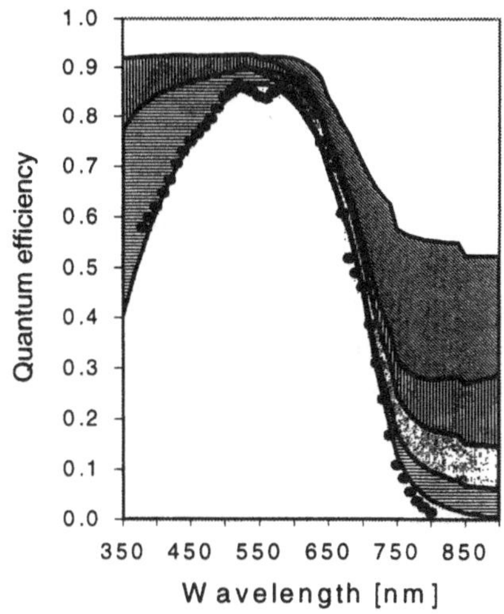

Figure 8.10 The measured (symbols) and calculated QE of the reference a-Si:H solar cell. The filling patterns are the same as in Fig. 8.8.

light that arrives at the back contact under a large angle. Taking into account the absorption in the TCO layer of photons that are incident at the interface at angles smaller than 30 degrees, the optimum thickness of ITO was 50 nm for the case of a textured a-Si:H solar cell. Implementing the optimized back contact in the cell resulted in a relative improvement in J_{sc} of 19 %.

8.3.2 Multijunction a-Si:H alloy solar cells

The advantage of a multijunction structure in a-Si:H solar cells is in using thin intrinsic layers with different optical band gaps. This leads to the absorption of a larger part of the solar spectrum and suppresses the effect of light-induced degradation as electric field in the intrinsic layers is higher. There are two crucial requirements when the multijunction structure is applied in a two terminal device: (i) the current generated at the maximum power point has to be equal in each component cell and (ii) the tunnel-recombination junction (TRJ) between the component cells has to feature low electrical and optical losses. Since the structure of the multijunction cells becomes increasingly complex, computer modeling is an important tool for analysis and understanding of these devices and their optimization.

The main difficulty in modeling multijunction solar cells are the n-p junctions comprising the TRJ between the adjacent component p-i-n cells. The TRJ serves as an electrical connection between the component cells in a two terminal device utilizing the internal carrier exchange effect. This effect is accomplished by recombination of electrons arriving to the TRJ from the top cell with holes from the bottom cell through the localized states at the interface of the junction. The carriers in the TRJ move to the interface of the TRJ and reach the recombination centers there by a tunneling mechanism. Appropriate models need to be developed and implemented that describe or approximate the processes at the TRJ. At present, two models have been presented in the literature for the TRJ that allow to simulate a multijunction solar cell as a single complete device and lead to realistic characteristics. These are the PennState model (Hou et al., 1991) and the Delft model (Willemen et al., 1994).

In order to avoid most difficulties in modeling the TRJ the most widespread and simplest approach for modeling multijunction solar cells is to simulate the component cells separately. The resulting J-V characteristic of the multijunction cell is then obtained by adding the J-V characteristics of individual component cells through series connection (Pawlikiewicz and Guha, 1990, Block, 1991). Pawlikiewicz and Guha, 1990 used this approach to compare the performance of tandem and triple junction amorphous silicon based solar cells. First, they demonstrated the validity of their modeling approach by achieving agreement between simulated and measured characteristics of in-house fabricated, 13.0 % tandem and 13.7 % triple cells (initial efficiency) with a minimum Tauc optical band gap of 1.4 eV (the state of the art in 1990). Based on the results of their simulations they predicted that when employing an even narrower band gap a-SiGe:H material of 1.3 eV, the triple junction device would exceed the tandem in efficiency by over 1 percent under global AM1.5 solar spectrum (14.2 % versus 13.2 %). With the assumption that both cells absorb the same amount of light, this difference is mainly due to the weaker electric field in the tandem cell in which the individual i-layer thicknesses are larger. It is interesting to note that Yang et al., 1997 reported a triple-junction amorphous silicon alloy solar cell with 14.6 % initial and 13.0 % stable conversion efficiencies which exceed the prediction of 1990. Recently, the initial efficiency was improved to 15.2 % Yang et al., 1998.

The above mentioned approach for modeling multijunction cells does not carry any information about the TRJ. In this approach the contacts of the individual cells are modeled as the ideal ohmic contacts and any possible electrical losses in TRJ are neglected. This approach cannot be used to determine the required properties of the TRJ leading to an optimized design. Hou et al., 1991 proposed an approach for modeling the tunnel junction in a-Si:H based multijunction solar cells so that the cells could be simulated as single structures.

They examined the functioning of the tunnel junction in triple junction a-Si:H based solar cells and outlined guidelines on how to make these contact regions. The approach of PennState University is based on introducing a highly defective layer ("x-layer") with reduced band gap at the n/p interface and grading the n^+-layer and p^+-layer in the regions adjacent to the defective layer. They concluded that the key process in the operation of these contact regions is recombination. They demonstrated that any material layer that enhances this recombination would enhance the cell performance if it did not strongly absorb light. Hence, such a layer could be a metal, a narrow band gap semiconductor, or a heavily defective semiconductor. They further showed that a proper band gap grading of n^+-layer and p^+-layer could enhance the contact effectiveness.

The PennState model is easily implemented in an a-Si:H device simulator provided that one can define graded layers within the device and properly model the DOS distribution in the region of the TRJ. Several groups have reported on using this approach in their simulations of tandem a-Si:H cells (Bruns et al., 1995, Zeman et al., 1997, Chatterjee and Palit, 1997). The drawback of the model is that there is no physical background for a strongly reduced band gap of the defective layer and no tunneling mechanism is taken into account. The grading of the doped layers is responsible for enhanced transport towards recombination centers in the defective layer. A new approach for modeling the TRJ in multijunction solar cells was presented by Willemen et al., 1994. This new approach is based on the Trap Assisted Tunneling model (Hurkx et al., 1992) and enhanced carrier transport in the high field region of the TRJ.

Zeman et al., 1997 demonstrated, that both with the PennState model and the Delft model for the TRJ, realistic J-V characteristics could be simulated for tandem cells. After calibrating the model parameters for the TRJ they obtained a good agreement between the simulated and measured J-V characteristics of an a-Si:H/a-Si:H tandem cell. They used the calibrated model to investigate the current matching in a-Si:H alloy tandem cells taking into account the optical enhancement by scattering at the rough interfaces of the cells (Zeman et al., 1997a). They investigated the effect of the ratio of the thicknesses of the top (i1) and bottom (i2) intrinsic layers on the external parameters of the tandem cells. Both for a-Si:H/a-Si:H and a-Si:H/a-SiGe:H tandem cells the simulations showed that there was no advantage in using thick layers in the bottom cell in order to increase the efficiency. For both types of tandem cells there is an optimum thickness i2, which in the case of a-Si:H/a-Si:H tandem cells is around 300 nm and in the case of a-Si:H/a-SiGe:H around 150 nm, using estimated stabilized dangling bond densities. This value depends on the scattering properties of the textured interfaces in the cell. The introduction of a-SiGe:H alloy with a Tauc optical gap of 1.5 eV in the bottom cell has

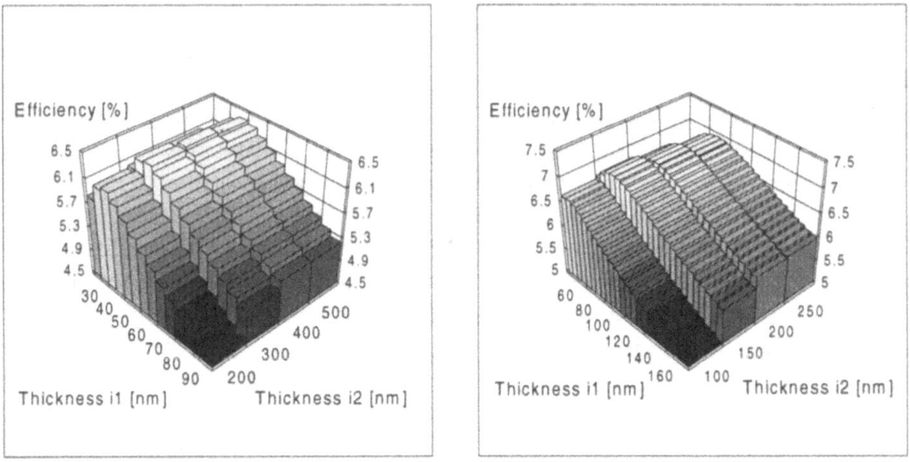

Figure 8.11. The efficiency of a-Si:H/a-Si:H (a) and a-Si:H/a-SiGe:H (b) tandem cells as a function of the thickness of the top (i1) and bottom (i2) intrinsic layers.

led to about 10 % increase in efficiency compared to a-Si:H/a-Si:H cells. Increased scattering in the cell leads to higher efficiencies and a lower ratio i2/i1 to obtain current matching, which can be achieved by making the top intrinsic layer thicker. This is desirable from a technological point of view, because the thicker top intrinsic layer helps to avoid the shunting problem and decreases the influence of pinholes. The efficiency of a-Si:H/a-Si:H and a-Si:H/a-SiGe:H tandem cells as a function of the thickness of the top (i1) and bottom (i2) intrinsic layers for a non-scattering structure is shown in Fig. 8.11.

With a device simulator for a multijunction cell as a single complete device including an appropriate model for the TRJ, one can investigate the interaction between the top and bottom cell. Bruns et al., 1995 used computer modeling to investigate the effect of current matching conditions on the spectral response characteristics of a tandem cell as a function of bias light and bias voltage. They found that the quantum efficiency is very sensitive to the current mismatch. Using the results of simulations he proposed an easy experimental approach for the determination of current matching. The device is biased to the desired operating conditions, for example with AM1.5 spectrum, and the quantum efficiency of the device is measured at the working point. The resulting quantum efficiency plot reveals the degree of current mismatch. If the matching is good the quantum efficiency is flat over the whole wavelength region. If the top cell

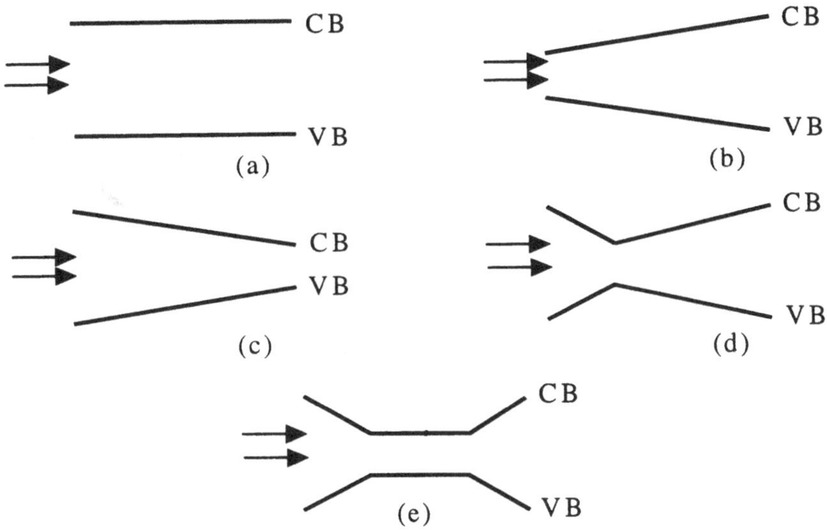

Figure 8.12. Band gap profiling configurations. (a) no profiling, (b) normal profiling, (c) reverse profiling, (d) double profiling or V shape, and (e) U shape.

limits the photocurrent, the quantum efficiency of the top cell will be found. In case of a limiting bottom cell, the plot shows the quantum efficiency of the bottom cell.

8.3.3 Band gap profiling of narrow band gap materials

An important step in obtaining higher efficiency in multijunction a-Si:H alloy solar cells was the introduction of band gap profiling of intrinsic layers in the component cells (Guha et al., 1989). Based on computer simulation studies Guha et al., 1989 demonstrated that with this novel structure for a given short circuit current, it was possible to obtain higher V_{oc} and FF than in the conventional cell design without profiling. Different types of profiling configurations, which are shown in Fig. 8.12 were investigated. In the case of normal profiling the band gap minimum is at the p-i interface through which the light enters the cell and increases linearly away from the interface. In the case of reverse profiling the band gap maximum is at the p-i interface and decreases away from it. Assuming that both the conduction and valence band edge positions shift equally as Ge is incorporated in the material, the computer simulations demonstrated that the FF is enhanced for normal profiling while V_{oc} is increased for reverse profiling. The computer simulations revealed that the double profiled structure (also called V shape profile) gave the best performance. The improvement in V_{oc} depends on the thickness of the region of reverse profiling. As the thickness increases, V_{oc} increases, however, *FF* starts decreasing at larger thicknesses since the holes have to move against a potential barrier over a larger distance. The results of Guha et al., 1989 were confirmed by more detailed computer analysis of V-shaped band gap grading in a-SiGe:H solar cells reported by Zimmer et al., 1997. The position of the band gap minimum was varied and its effect on the performance of the cell was investigated and compared with the experimentally fabricated cells. The results of the simulations corresponded well with the measured external parameters of the solar cells. The overall generation rate within the cell was found nearly independent of the position of the band gap minimum. The optimal position of the band gap minimum was determined to be close to the p-i interface. By analyzing the internal parameters of the cell such as the internal electric field, the R-G rate, and charge occupation of the localized states in the cell at different light and voltage bias conditions, more insight into transport and recombination behavior within the solar cell was gained. The simulations revealed that the position and the charge state of the defects determined the device characteristics. The simulation results pointed out that when using U-shaped grading (Fig. 8.12e) rather than V-shaped grading, together with suitably designed p/i and i/n buffer layers the performance of a-SiGe:H solar cell can be further improved (Vasanth et al., 1995). Later, it was shown (Fölsch et al., 1997) that, also in the degraded state, U-shaped grading in the a-SiGe:H component cell gives better performance than a V-shaped profile.

References

Arch, J.K., F.A. Rubinelli, J.Y. Hou, and S.J. Fonash, *First principles computer model showing the effect of p-layer thickness and front contact barrier height on the performance of a-Si:H p-i-n solar cells*, Proc. of the 21st IEEE PVSC (1990) 1636-1641.

Block, M., D. Bonnet, and F. Zetche, *Modeling of multilayer amorphous thin film silicon-germanium single and tandem solar cells*, Proc. of 22nd IEEE PVSC, (1991) 1275-1280.

Bruns, J., M. Choudhury, H.G. Wagemann, *The influence of working conditions on the current matching in a-Si:H stacked solar cells*, Proc. of the Thirteenth European Photovoltaic Solar Energy Conference, Nice, France, 23-27 October 1995, 230-233.

Chatterjee, P., *Photovoltaic performance of a-Si:H homojunction p-i-n solar cells: A computer simulation study*, J. Appl. Phys. **76** (1994) 1301.

Chatterjee, P., *A computer analysis of the effect of a wide-band-gap emitter layer on the performance of a-Si:H-based heterojunction solar cells*, J. Appl Phys. **79** (1996) 7339-7347.

Chatterjee, P., and N. Palit, *Numerical optimisation of double junction solar cells with a-Si:H absorber layers*, Proc. of the Fourteenth European Photovoltaic Solar Energy Conference, Barcelona, Spain, June 30 - July 4, 1997, 517-520.

Fölsch, J., D. Lundszien, F. Finger, H. Stiebig, J. Zimmer, C. Beneking, S. Wieder, and H. Wagner, *Stability investigation of a-Si:H/a-SiGe:H tandem solar cells*, Proc. of the Fourteenth European Photovoltaic Solar Energy Conference, Barcelona, Spain, June 30 - July 4, 1997, 601-604.

Guha, S., J. Yang, A. Pawlikiewicz, T. Glatfelter, R. Ross, and S.R. Ovshinsky, *Band-gap profiling for improving the efficiency of amorphous silicon alloy solar cells*, Appl Phys. Lett. **54** (1989) 2330-2332.

Hack, M. and M. Shur, *Physics of amorphous silicon alloy p-i-n solar cells*, J. Appl. Phys. **58** (1985) 997-1020.

Hishikawa, Y., E. Maruyama, S. Yata, M. Tanaka, S. Kiyama, and S. Tsuda, *Optical Confinement in High Efficiency a-Si Solar Cells with textured Surfaces*, Technical Digest of the International PVSEC-9, Miyazaki, Japan, 1996, 639-640.

Hou, J.Y., J.K. Arch, S.J. Fonash, S. Wiedeman, and M. Bennet, *An examination of the "tunneljunctions" in triple junction a-Si:H based solar cells: Modeling and effects on performance*, Proc. 22nd IEEE PV Specialists Conference, Las Vegas (1991) 1260-1264.

Hurkx, G.A.M., D.B.M. Klaassen and M.P.G. Knuvers, *A new recombination model for device simulation including tunneling*, IEEE Transactions on Electron Devices **39** (1992) 331-338.

Leblanc, F., J. Perrin, and J. Schmitt, *Numerical modeling of the optical properties of hydrogenated amorphous-silicon-based p-i-n solar cells deposited on rough transparent conducting oxide substrates*, J. Appl. Phys. **75** (1994) 1074-1086.

McElheny, P.J., P. Chatterjee, and S.J. Fonash, *Collection efficiency of a-Si:H Schottky barriers: A computer study of the sensitivity to material and device parameters*, J. Appl. Phys. **69** (1991) 7674-7688.

Meyer, R.R. and P.M. Roth, *Modified Damped Least Squares: an algorithm for non-linear optimization*, Journal of the Institute of Mathematics and its Applications **9** (1972) 218-233.

Morris, J., R.R. Arya, J.G. O'Dowd, and S. Wiedeman, *Absorption enhancement in hydrogenated amorphous silicon-based solar cells*, J. Appl. Phys. **67** (1990) 1079-1087.

Ouwerling, G.J.L., *The Profile/Prof2d User's Manual*, Delft University of Technology, 1987.

Pawlikiewicz, A.H., and S. Guha, *Performance comparison of triple and tandem multi-junction a-Si:H solar cells: A numerical study*, IEEE Transactions on Electron Devices **37** (1990) 1758-1762.

Smole, F., and J. Furlan, *Effects of abrupt and graded a-Si:C:H/a-Si:H interface on internal properties and external characteristics of p-i-n solar cells*, J. Appl. Phys. **72** (1992) 5964-5969.

Smole, F., M. Topič, J. Furlan, *Amorphous silicon solar cell computer model incorporating the effects of TCO/a-Si:C:H junction*, Solar Energy Materials and Solar Cells **34** (1994) 385-392.

Stiebig, H., A. Kreisel, K. Winz, N. Schultz, C. Beneking, Th. Eickhoff and H. Wagner, *Spectral response modeling of a-Si:H solar cells using accurate light absorption profiles*, 1st World Conference on Photovoltaic Energy Conversion, (Proc. 24th IEEE PV Specialists Conference, Waikoloa, HI, USA, December 1994) 603-606.

Tao, G., B.S. Girwar, G.E.N. Landweer, M. Zeman, and J.W. Metselaar, *Highly reflective TCO/Al back contact for a-Si:H solar cells*, Proc. of the 11th European Photovoltaic Solar Energy Conference, Montreux (1992) 605-608.

Tasaki, H., W.Y. Kim, M. Hallerdt, M. Konagai, and K. Takahashi, *Computer simulation model of the effects of interface states on high performance amorphous silicon solar cells*, J. Appl. Phys. **63** (1988) 550-560.

Vasanth, K., A. Payne, B. Crone, S. Sherman, M. Jakubowski and S. Wagner, *Design and growth of band-gap graded a-SiGe:H solar cells*, in: Amorphous Silicon Technology - 1995, edited by M. Hack, E.A. Schiff, A. Madan, M. Powell, and A. Matsuda, Materials Research Society Symp. Proc. **377** (1995) 663-668.

Willemen, J.A., M. Zeman, and J.W. Metselaar, *Computer modeling of amorphous silicon tandem cells*, 1st World Conference on Photovoltaic Energy Conversion, (Proc. 24th IEEE PV Specialists Conference, Waikoloa, HI, USA, December 1994) 599-602.

Willemen, J.A., M. Zeman and J.W. Metselaar, *Sensitivity study of a-Si:H solar cell model parameters*. Technical Digest of the International PVSEC-9. Miyazaki, Japan, (1996) 359-360.

Willemen, J., *Modelling of amorphous silicon single and multi-junction solar cells*, PhD thesis, Delft University of Technology, 1998.

Yang, J., A. Banerjee, and S. Guha, *Triple-junction amorphous silicon alloy solar cell with 14.6 % initial and 13.0 % stable conversion efficiencies*, Appl. Phys. Lett. **70** (1997) 2975-2977.

Yang, J., A. Banerjee, S. Sugiyama, and S. Guha, *Correlation of component cells with high efficiency amorphous silicon alloy triple-junction solar cells and modules*, presented at the 2nd World Conference and Exhibition on Photovoltaic Energy Conversion, to be published.

Zeman, M., J.A. Willemen, S.Solntsev, J.W. Metselaar, *Extraction of amorphous silicon solar cell parameters by inverse modeling*, Solar Energy Materials and Solar Cells **34** (1994) 557-563.

Zeman, M., J.A. Willemen, L.L.A. Vosteen, G. Tao and J.W. Metselaar, *Computer modeling of current matching in a-Si:H/a-Si:H tandem solar cells on textured substrates*, Solar Energy Materials and Solar Cells **46** (1997) 81-99.

Zeman, M., R.E.I. Schropp, and W. Metselaar, *Computer modeling of a-Si:H alloy tandem cells: Determination of current matching with light scattering*, Proceedings of the Fourteenth European Photovoltaic Solar Energy Conference, Barcelona, Spain, June 30 - July 4, 1997, 586.

Zeman, M., R.A.C.M.M. van Swaaij, E. Schroten, L.L.A. Vosteen, and J.W. Metselaar, *Device modeling of a-Si:H alloy solar cells: Calibration procedure for determination of model input parameters*, in: Amorphous and Microcrystalline Silicon Technology - 1998, edited by R. Schropp, H. Branz, S. Wagner, M. Hack, and I. Shimizu, Materials Research Society Symp. Proc. **507** (1998) in print.

Zhu, H., and S.J. Fonash, *Band-offset determination for intrinsic a-Si/p+ μc-Si or intrinsic a-Si/n+ μc-Si heterostructure*, in: Amorphous and Microcrystalline Silicon Technology - 1998, edited by R. Schropp, H. Branz, S. Wagner, M. Hack, and I. Shimizu, Materials Research Society Symp. Proc. **507** (1998) in print.

Zimmer, J., H. Stiebig, J. Fölsch, F. Finger, Th. Eickhoff and H. Wagner, *More insight into band gap graded a-SiGe:H solar cells by experimental and simulated data*, in: Amorphous and Microcrystalline Silicon Technology - 1997, edited by S. Wagner, M. Hack, E.A. Schiff, R. Schropp, and I. Shimizu, Materials Research Society Symp. Proc. **467** (1997) 735-740.

About the Authors

Ruud E.I. Schropp was born in Maastricht, the Netherlands, on September 20, 1959. He obtained his Ph.D. degree in 1987 from Groningen State University, the Netherlands, on his research on thin film transistors (TFTs) made of amorphous silicon. From 1987 till 1989 he was with Glasstech Solar, Inc., in Colorado, U.S.A., where he was manager R & D and responsible for optimizing deposition equipment and processing parameters for the fabrication of amorphous silicon solar cells in the laboratory as well as in production. From 1989 till 1994 he held a research fellow position at Utrecht University, the Netherlands, where he initiated a number of projects related to photovoltaics. From 1994 onward he has been a faculty senior researcher and lecturer, thus promoting the field of thin film deposited materials and devices. In 1996 and 1997 he was co-organizer and in 1998 organizer of the *Amorphous and Microcrystalline Silicon Technology* symposium of the Materials Research Society.

Miro Zeman was born in Žilina, Slovakia, on June 21, 1957. He received his Ph.D. degree in 1989 from the University of Technology in Bratislava, Slovakia, for his work on the fabrication of devices for electronics based on amorphous silicon material. In the same year he joined the solar cell group at Delft University of Technology in the Netherlands, where he worked as a research fellow on the implementation of amorphous silicon-germanium alloys in multijunction amorphous silicon solar cells and on optimizing the solar cell structures by using computer modeling. From 1995 onward he has been a senior researcher at the Delft Institute of Microelectronics and Submicron Technology (DIMES) at Delft University of Technology, where he has been in charge of a number of projects related to thin film solar cells. In 1998 he has started lecturing at courses on photovoltaic conversion.

Index

FSC
www.fsc.org
MIX
Papier | Fördert gute Waldnutzung
FSC® C083411

Kolibri 360 GmbH
Ferdinand-Jühlke-Straße 7
99095 Erfurt, Deutschland
produktsicherheit@kolibri360.de